KINEMATICS ANALYSIS AND SYNTHESIS

McGRAW-HILL SERIES IN MECHANICAL ENGINEERING

Consulting Editors

Jack P. Holman, *Southern Methodist University*
John R. Lloyd, *Michigan State University*

Anderson: *Modern Compressible Flow: With Historical Perspective*
Arora: *Introduction to Optimum Design*
Bray and Stanley: *Nondestructive Evaluation: A Tool for Design, Manufacturing and Service*
Culp: *Principles of Energy Conversion*
Dally: *Packaging of Electronic Systems: A Mechanical Engineering Approach*
Dieter: *Engineering Design: A Materials and Processing Approach*
Eckert and Drake: *Analysis of Heat and Mass Transfer*
Edwards and McKee: *Fundamentals of Mechanical Component Design*
Heywood: *Internal Combustion Engine Fundamentals*
Hinze: *Turbulence*
Hutton: *Applied Mechanical Vibrations*
Juvinall: *Engineering Considerations of Stress, Strain, and Strength*
Kays and Crawford: *Convective Heat and Mass Transfer*
Kane & Levinson: *Dynamics: Theory and Applications*
Kimbrell: *Kinematics Analysis and Synthesis*
Martin: *Kinematics and Dynamics of Machines*
Phelan: *Fundamentals of Mechanical Design*
Raven: *Automatic Control Engineering*
Rosenberg & Karnopp: *Introduction to Physics*
Schlichting: *Boundary-Layer Theory*
Shames: *Mechanics of Fluids*
Sherman: *Viscous Flow*
Shigley: *Kinematic Analysis of Mechanisms*
Shigley and Uicker: *Theory of Machines and Mechanisms*
Shigley and Mischke: *Mechanical Engineering Design*
Stoecker and Jones: *Refrigeration and Air Conditioning*
Vanderplaats: *Numerical Optimization: Techniques for Engineering Design, with Applications*
White: *Viscous Fluid Flow*

KINEMATICS
ANALYSIS
AND
SYNTHESIS

Jack T. Kimbrell

Professor Emeritus
of Mechanical Engineering
Washington State University

McGraw-Hill, Inc.

New York   St. Louis   San Francisco   Auckland   Bogotá   Caracas
Hamburg   Lisbon   London   Madrid   Mexico   Milan   Montreal   New Delhi
Paris   San Juan   São Paulo   Singapore   Sydney   Tokyo   Toronto

This book was set in Times Roman.
The editors were John J. Corrigan and David A. Damstra;
the production supervisor was Leroy A. Young.
The cover was designed by Joseph Gillians.
R. R. Donnelley & Sons Company was printer and binder.

KINEMATICS ANALYSIS AND SYNTHESIS

1 2 3 4 5 6 7 8 9 0 DOC DOC 9 0 9 8 7 6 5 4 3 2 1

ISBN 0-07-034576-7

Library of Congress Cataloging-in-Publication Data

Kimbrell, Jack T.
    Kinematics analysis and synthesis / Jack T. Kimbrell.
       p.   cm.
    Includes bibliographical references.
    ISBN 0-07-034576-7
    1. Machinery, Kinematics of.   2. Machinery, Dynamics of.
I. Title.
TJ175.K56   1991
621.8'11—dc20   90-42290

ABOUT THE AUTHOR

Jack T. Kimbrell is Professor Emeritus and former Chairman of Mechanical Engineering and Acting Dean at Washington State University. He has taught and conducted research since 1946 at the University of Missouri, Midwest Research Institute, and Washington State University.

He has served as a design consultant to national and international companies and is currently president of his engineering consulting firm in New Mexico. He is a member of the American Society of Mechanical Engineers and a former Alcoa Professor of Design.

Professor Kimbrell is a graduate of Purdue University and the University of Missouri and is a registered Professional Engineer in Washington State and New Mexico.

CONTENTS

PREFACE

The design of any new machine or the analysis of an existing machine generally starts with the kinematics that are involved. After the existing or proposed motions are analyzed for velocity, acceleration, and static forces, it is possible for the designer to consider strength, mass, inertia forces, and dynamic balancing. Kinematic analysis does not consider strength of the members but only concerns itself with motions involved. Thus the study of kinematics is a vital element of the machinery design process.

Analysis is done either graphically or analytically depending upon the experience and/or desires of the designer. A designer well experienced in kinematic analysis may opt to do analytic analysis, while the novice will gain greater insight using the graphical procedures. Often it is desirable to construct a simple paper or wood model of the mechanism in order to better visualize the motions or to evaluate space requirements.

The ever-increasing demand for automation in industry has caused more emphasis to be placed on kinematic analysis and design of mechanisms. A natural extension of the planar mechanisms considered in this text is that of robotic systems. Many favor the three-dimensional robotic systems, while two or more planar systems with coordinated motion may perform more dependably. The "pick and place" robotic system is essentially two coordinated planar systems.

The text is arranged in a systematic order starting with simple motion analysis and proceeding through velocity and acceleration considerations. The traditional graphical techniques are explained in detail. Following the graphical techniques the analytic techniques are explained. The presentation rather than the mathematical niceties is stressed. An energetic student should be able to follow the analytic procedures and even bring some original thinking to the process.

With the increasing capabilities of personal computers and their adoption in industry, computer programs written in GW BASIC are included in the appendix. The GW BASIC language was selected since it is a popular programming language found with most personal computers. The programs are written in rather simple format without taking advantage of many step-saving procedures provided by the language. In this manner, the student should have little or no trouble adapting the programs to his or her particular computer. It is expected that the student can shorten the programs by utilizing the string functions and data input procedures that may be available with many BASIC programming units.

Since kinematics and the geometry of motions are very closely interwoven, a chapter is devoted to the geometry of motion. It is expected that an understanding of the geometry involved will provide the student with a much firmer foundation for advanced study of kinematics.

The section on cams and cam design can be viewed as another means of providing programmed motion to a system. The constant velocity, constant acceleration, simple harmonic, and cycloidal cam follower motions are discussed in detail. A computer program is provided to aid in analysis of a system using simple harmonic or cycloidal follower motion. Handling the dynamics problem of cam-driven mechanisms should be a natural consequence of the application of the previous chapters on velocity, acceleration, and acceleration forces.

Geometry and notation involved in spur, helical, bevel, and worm gears is included. The text uses standard notation currently recommended by the American Gear Manufacturers Association (AGMA) and also that recommended by the International Standards Organization (ISO) for metric gears. Examples are provided for analysis or design using the AGMA or the ISO units.

Mathematical abilities commonly found in the fourth semester of college study are assumed. Some knowledge of vectors and vector notation is desirable. An awareness of linear algebra is helpful but not essential. Since the equations developed are all linear, a computer program for solution of simultaneous linear equations is included in the appendix. Although matrix manipulations are not necessary, some may prefer that technique, and a short review of matrix manipulations is included in Appendix B along with a computer program for matrix inversion.

For design of mechanisms to satisfy up to four specified positions, the equations are all linear. For design of four-bar mechanisms to satisfy more than four positions, the least-squares technique results in linear equations. In order to design a generalized mechanism with more than four bars, the equations become nonlinear and the Newton-Raphson technique is recommended. The bibliography includes references which present the theory of the Newton-Raphson technique and sources for computer programs. An example is included along with a BASIC program using the Newton-Raphson method.

Other than the section dealing with synthesis using the instant center for acceleration the author does not claim originality in the topics considered. Many very informative references are provided by subject in the bibliography.

ACKNOWLEDGMENTS

I would like to thank the following reviewers for their many useful comments and suggestions: Dan Afolabi, Indiana University; Aly El-Shafei, University of South Florida; Edward R. Garner, California Polytechnic San Louis Obispo; Richard Golembiewski, Milwaukee School of Engineering; Jordan L. Larson, Iowa State University; Robert A. Lucas, Lehigh University; Levern A. Reis, Texas Tech University; Willem Stuiver, University of Hawaii; Wiesliaw Szydlowski, University of Houston; Lung-Wen Tsai, University of Maryland; and my wife, Maxine, for her inspirational encouragement and infinite patience.

Jack T. Kimbrell

KINEMATICS ANALYSIS AND SYNTHESIS

CHAPTER
1

MECHANISMS
ANALYSIS
AND
SYNTHESIS

1.1 INTRODUCTION

The design engineer plays a significant role in mechanization and automation of modern industry. The designer is called upon to devise a machine or system of machines which is capable more often than not of accepting and following instructions emanating from a computer, performing those required tasks reliably, and notifying the computer when the task is complete. The designer must be familiar with and consider such factors as stress, strength, flexibility, dynamic balance, material properties, manufacturing influences, safety, costs, noise, codes and standards, efficiency, aesthetics, wear, lubrication, maintenance, deflections, accuracy, and reliability as well as control systems and a host of other influencing elements.

In the face of all that must be considered, it is often more effective to set aside many of the influencing factors and reduce the problem to a more readily solved series of smaller tasks. Other influences are then recalled individually and the problem solution modified accordingly. Unfortunately, the final solution inevitably represents a series of necessary compromises.

The study of mechanisms removes all influencing factors except those concerned with position and time. As such it represents a study of motion and geometry aimed at creating a mechanism which will satisfactorily meet the requirements of the problem at hand. After the mechanism is devised, the concepts of mass and inertia forces are introduced to allow thoughts of dynamic balance, shaking forces, and noise. This procedure has the effect of replacing empiricism and practicality with theory.

Although there is much to be said in favor of empirical and practical practice tempered with experience, it is difficult to imagine effective treatment of new and unsolved problems without strong reliance on theory. It is essential that the theoretical basis be thoroughly understood before the empirical ideas are used to temper the results. This is not to belittle the experience factor but simply to provide a sound basis for decision making. It is the engineering method in practice.

1.2 HISTORICAL BACKGROUND

The study of mechanisms began in antiquity under the pressure of necessity. At the outset there was neither a plan nor unity but rather a random growth resulting in an amazing assemblage of mechanisms. The many wars throughout history resulted in development of such items as "throwing machines" for throwing larger rocks greater distances and progressing through rather complex guidance systems for missiles. If any good at all has resulted from wars, it must include advancement of the study of mechanisms.

Ancient rulers and their whims were effective in promoting the study of mechanisms. A classic example is the Marly machine devised for the sole purpose of creating beautiful fountains in the garden of the palace at Versailles. The Marly machine utilized fourteen 40-foot(ft)-diameter undershot water wheels at the river to develop 1200 horsepower(hp). The water wheels drove 64 pumps at the river, 79 pumps at a 160-ft elevation, and 82 pumps at a 343-ft elevation. Water was delivered to an aqueduct at 533 ft above the river. All the pumps were driven with parallelograph linkages which required 64,000 ft of iron bars.

Commercial necessity also prompted mechanical innovation. The need to move merchandise resulted in some fascinating mechanisms such as the paddle wheel of Buchannan (Fig. 1.1) arranged to keep the paddles perpendicular to the water stream and Watts' steam engine (Fig. 1.2).

Perhaps the earliest writings on the subject were those of Pollio Marcus Vitruvius, a military engineer who wrote *De Architectura* in about 28 B.C. Vitruvius was principally concerned with methods for moving heavy objects. It was Hero of Alexandria in about the first century A.D., essentially a mathematician but enamored with applications, who named the components from which all complex mechanisms of that day were

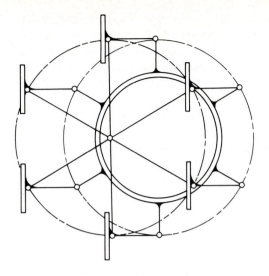

FIGURE 1.1
Buchannan's paddle wheel.

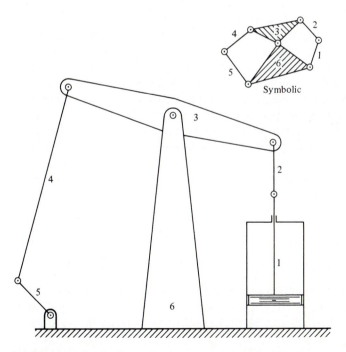

FIGURE 1.2
Watts' steam engine.

assembled. He recognized the wedge, lever, screw, windlass, and pulley. In these early times machines were considered as whole objects, not as groups of parts known today as mechanisms. In 1588, Ramelli in *Arteficiose Machine* describes each machine separately and completely without any recognition of similarity of the machine components. Jacob Leupold appears to have been the first to recognize mechanisms as components of machines and to suggest the idea of modifying motion. His nine-volume series of books published between 1724 and 1739 was directed toward craftspeople. The books contained almost 500 copper engravings depicting the then current practice in machines. The illustrations were so complete that any competent craftsperson could reproduce the machines and mechanisms.

It was the work of André Marie Ampère (1775–1836) that stimulated the science of kinematics as a separate study. During his critical examination of mechanisms at Ecole Polytechnique, Ampère published his *Essai sur la Philosophie de Science*. He gave the name kinematics (*Cinématique*) to the study and encouraged separate treatment of the subject. Following the lead of Ampère, the science of kinematics was taken up as a separate study and continues today supported by a wealth of literature.

Prior to about 1940, significant literature on the subject of kinematics originated in Europe and Australia. The literature followed either a discussion of graphical techniques (applied geometry) or highly analytic methods. The bulk of the writings was devoted to plane mechanisms. In 1948 the Mechanisms Conference organized at Purdue University sparked an interest in the subject in the United States, and significant literature has resulted. With the advent of computers, the literature expanded from analysis of plane mechanisms to analysis and synthesis of plane and spatial mechanisms. The applications to robots are obvious, and computer-controlled automated systems using plane and spatial mechanisms are a very natural development. Unfortunately, very little has been accomplished in recent years in extending or developing new theories of kinematics. Major efforts in the United States and abroad have been in application of existing theories. It is doubtful though that all theories possible have been developed.

1.3 SCHEMATIC NOTATION

In considering time-dependent motion of machine parts, it is unnecessary, and in fact complicating, to be concerned with the actual shape and mass of the parts. For this reason, several simplifying conventions have been adopted.

A mechanism consists of several component parts which move relative to one another. Each component part is called a *link*. When two links are connected together in some manner which allows relative motion, the connection is known as a *pairing element* or a *joint*. Thus, several links connected by pairing elements constitute a *chain*. A chain is frequently recognized by the number of links which make it up. Such recognition results in calling the links *bars*. A four-bar mechanism consists of four links connected by pairing elements in such a manner as to allow relative motion between the links.

Each link of a mechanism is assumed (at the outset) to be totally rigid. It will neither stretch, compress, bend, or twist but will always be the same size regardless of the motion involved. This is known to be false, but the error involved is very small and the assumption is justified because of the resulting simplicity.

A mechanism link is assumed to have zero thickness perpendicular to its plane of motion. With this assumption, an infinite number of links may occupy the same space in the plane of operation without any interference with one another's motion. After the kinematic analysis of motion and before the mechanism is constructed, the designer must remember that this assumption was made and must arrange the final design so that interference will not occur.

Knowing the location of pairing elements at all times is very important in kinematic analysis. It is necessary, at the outset, to locate those elements on a sketch without undue complication of the sketch. The connecting rod of an internal combustion engine may look somewhat like that shown in Fig. 1.3a. Since, for kinematic analysis, the significant components of the connecting rod are the pairing elements where the connecting rod is attached to the crank shaft and piston, a connecting rod is shown kinematically in Fig. 1.3b.

The true shape of the connecting rod is not obvious in the kinematic drawing, and an additional assumption is made to the effect that the link is infinite in area in the plane of motion. That is, any point in the plane of

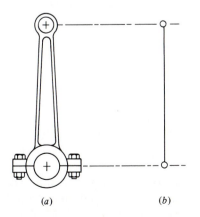

(a)   (b)

FIGURE 1.3
Kinematic representation.

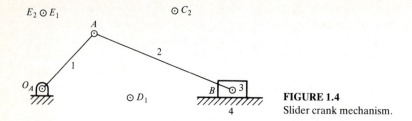

FIGURE 1.4
Slider crank mechanism.

motion of the connecting rod may be considered as a point on the connecting rod whether or not it appears on the line joining pairing elements.

A kinematic drawing of an internal combustion engine (known as a slider crank mechanism which is a four-bar mechanism) is shown in Fig. 1.4.

In Fig. 1.4, the link numbered 2 is the connecting rod which connects the crank shaft (link 1) with the piston (link 3). Link 4, designated by the sloping short lines, is the frame or block of the engine. The sloping short lines indicate a stationary (nonmoving) link of the mechanism. Point O_A is that fixed point on the stationary link about which link 1 rotates (the engine main bearing). Link 3 is restricted to slide on link 4 and will always be in contact with link 4.

Point C_2 is known to be a point on the connecting rod because of the subscript 2. In like manner, point D_1 is a point known to be on the crank shaft. Its motion is rotation with point O_A as the center of rotation. Note that point E_2 on the connecting rod and point E_1 on the crank shaft occupy the same space. This does not mean that they are in any way connected to one another. They simply happen to be at the same point in the plane and may be moving in different directions with different velocities at this particular instant. At point A, links 2 and 3 are connected by a pairing element. Since A_2 and A_3 are connected, these two points will always be moving in the same direction with the same velocity and acceleration.

In some instances numerical subscripts refer to various positions of the points. When this occurs, it is generally very obvious and no confusion results.

1.4 CLASSIFICATION OF CHAINS

Kinematic chains are classified in accordance with the type of motion they produce. Four-bar mechanisms can be one of three types of chains depending on how the links move.

The *crank rocker* mechanism is one in which the input crank will make a 360° rotation about its center of rotation and the output link will only oscillate and not rotate 360° (Fig. 1.5a).

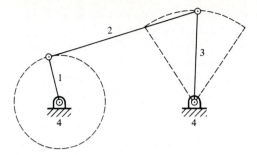

(*a*) Crank rocker

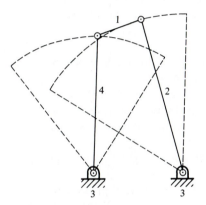

(*b*) Double rocker

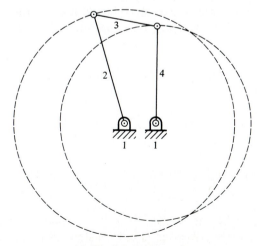

(*c*) Drag link (double crank)

FIGURE 1.5
Grashof mechanisms.

The *double rocker* mechanism is one in which neither of the links will make a 360° rotation but will only oscillate (Fig. 1.5*b*).

The *drag link* mechanism is one in which two of the links will make a 360° rotation about their centers of rotation (Fig. 1.5*c*). The drag link mechanism is sometimes referred to as a *double crank* mechanism.

The Grashof (1883) criterion may be used to recognize various types of frame-connected four-bar mechanisms. A frame-connected mechanism is one in which one of the links is made stationary and is considered to be the mechanism base. The Grashof criterion is based on the length of the four links which make up the mechanism. If the sum of the length of the longest and shortest links of a mechanism is *less* than the sum of the lengths of the other two links, then a Grashof mechanism results.

If the shortest link of a Grashof mechanism serves as the crank and an adjacent link as the base, two different crank rocker mechanisms may be recognized (Fig. 1.6).

If the shortest link of a Grashof mechanism is aligned opposite the base link, then only a double rocker mechanism can exist (Fig. 1.5*b*).

If the shortest link of a Grashof mechanism serves as the base, then only a drag link or double crank mechanism can exist (Fig. 1.5*c*).

If the sum of the length of the longest and shortest links is greater than the sum of the lengths of the other two links, then only double rocker mechanisms can be assembled regardless of which link is made the base.

If the sum of the length of the longest and shortest links is equal to the sum of the lengths of the other two links, then all three types of mechanisms are possible as in the case of the sum of the longest and shortest being less than the sum of the length of the other two links. However, each mechanism will suffer from change points. When change points exist, the centerlines of each link will become collinear and, unless guidance is provided, the cranks will be able to change direction of rotation.

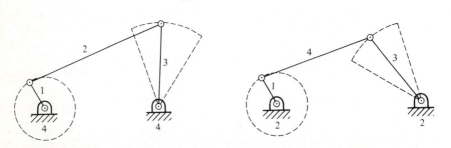

FIGURE 1.6
Alternate crank rocker mechanisms.

A linkage with all four links of the same length is known as a parallelogram linkage which can only be a double crank mechanism if it is guided through the change points.

The Galloway mechanism is a special case in which the sum of the lengths is equal. It is a deltoid linkage with two short links equal in length connected to two longer links equal in length. With one of the short links as the base, a double crank mechanism results with the short rotating link making two revolutions to one revolution of a longer link. With one of the long links as the base, a crank rocker mechanism is possible.

1.5 EXPANSION AND INVERSION

Expansion is a natural consequence of one of the assumptions associated with schematic notation. Since each link of a mechanism is represented only by a straight line connecting each of the pairing elements, the true shape of the link is not known. Therefore, it is possible to believe that any point in the plane of the link drawing is considered to be a point on that link. It is not essential nor is it relevent whether the point is on the line in the drawing joining the pairing elements. With this thought firmly in mind one must believe that any point in the plane of the link may be considered as a point on the link. In like manner, any point in the plane of motion may be considered to be a point on any link in the mechanism. Thus, two points which occupy the same position in the plane of motion may be considered as independent points associated with two different links of the mechanism.

Inversion refers to the process of considering the fixed (base) link to be free to move and another link of the mechanism to be the fixed (base) link. This is a process which results in a completely different mechanism made up of the same links as the original mechanism. However, since the lengths of the links do not change, the relative rotation between the links is unchanged.

The inversion technique provides a convenient means for analysis of a mechanism and is particularly helpful in synthesis of mechanisms. Figure 1.7 shows a crank rocker mechanism and its inverted configurations. Note, in Fig. 7, that if the angle α between links 1 and 2 becomes 30° smaller, then in all cases the angle β between links 3 and 4 becomes 6° smaller. Figure 1.8 shows a slider crank mechanism and its inverted configurations. Note that if the angle α is increased the same amount in each inversion, the distance x remains the same regardless of which inversion is considered. The significant feature of the inversion technique is that the relative motion between the links is unchanged by the inversion process. This relationship is of importance in analysis of complex mechanisms and forms the basis of the ingenious method of Goodman for indirect acceleration analysis.

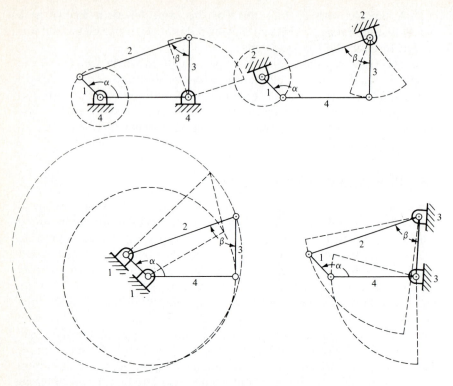

FIGURE 1.7
Inversions of a crank rocker mechanism.

1.6 PAIRING ELEMENTS

Pairing elements are the connections between the links of a kinematic chain. They permit relative motion between the links, but they serve to connect each link with another. In Fig. 1.9, the pairing elements that connect links 1 and 2, 2 and 3, and 1 and 4 allow only relative rotation between the connected links. Such pairing elements are known as *revolute* joints. The connection between links 4 and 3 is a pairing element which allows only sliding motion between links 4 and 3. Such a pairing element is known as a *prismatic* joint.

If a pairing element allows relative motion between connected links which can be described by only one quantity, that pairing element is known as a *lower pair*. Note that all the pairing elements shown in Fig. 1.9 are lower pairs. In the event that two quantities are necessary to describe the relative motion between two connected links, that pairing element is known as a *higher pair*.

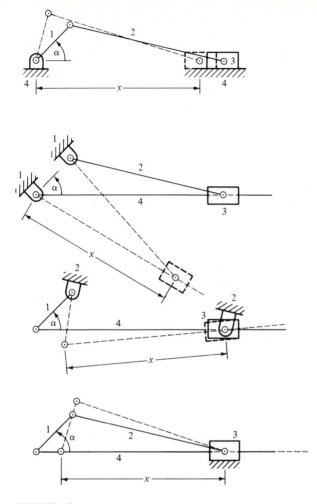

FIGURE 1.8
Inversions of a slider crank mechanism.

Table 1.1 shows those pairing elements along with their schematic representation and the number of quantities required to describe the relative motion between connected links.

1.7 MOBILITY (Degrees of Freedom)

A single link with plane motion (moves in one plane only) will have three ($F = 3$) degrees of freedom of motion. An assemblage of links (a chain or a mechanism) with n links would be expected to have a total of $3n$

TABLE 1.1
Pairing elements

Type	Schematic	Motion	Describing quantity	Degree of freedom
Revolute		Rotation	θ	1
Prismatic		Sliding	x	1
Helical		Rotation and translation	θ or x	1
Rolling (No slip) (With slip)		Rotation and translation	θ or x / θ and x	1 / 2
Cylindric		Rotation and sliding	θ and x	2
Planar		x, y translation and rotation about z axis	x, y, and θ	3
Global		Rotation about x, y, and z axes	θ, λ, and ϕ	3

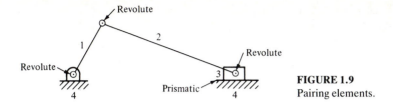

FIGURE 1.9
Pairing elements.

degrees of freedom of motion. However, since one link of a mechanism is considered to be the base and is fixed so that it cannot move, the degrees of freedom of motion is reduced to $3(n - 1)$. In addition, the pairing elements reduce the freedom of motion depending on the type of pairing element. Thus the total degrees of freedom of a plane motion mechanism may be expressed as

$$F = 3(n - 1) - 2L - H$$

where F = total degrees of freedom of mechanism
n = total number of links of mechanism
L = total number of lower pairs
H = total number of higher pairs

A mechanism with zero degrees of freedom ($F = 0$) will not have any relative motion between the links and is considered to be a rigid frame. Rigid frames are used in the design of bridges, buildings, aerodynamic structures, etc. A rigid frame is not very useful in providing motion as is normally expected from a mechanism.

A mechanism with one input drive system and with one degree of freedom ($F = 1$) will perform in a predictable manner, and it will be possible to analyze the mechanism for motion, velocity, acceleration, inertial forces, and dynamic balance.

The Whitworth quick-return mechanism shown in Fig. 1.10 has six links and seven lower pairs (five revolute and two prismatic pairs) which yield one degree of freedom.

$$F = 3(6 - 1) - 2(7) = 1$$

A mechanism with two input drive systems and with two degrees of freedom ($F = 2$) will perform in a predictable manner. However, if a mechanism has two degrees of freedom and only one input drive system, that mechanism will have unpredictable motion. Some consider such mechanisms useless, but these mechanisms are successfully used in children's toys.

In general, a mechanism must have an equal number of input drive systems and degrees of freedom before it will perform in a predictable manner.

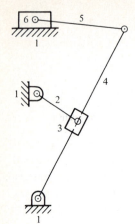

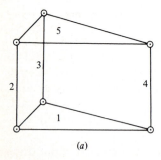

FIGURE 1.10
Whitworth mechanism.

In analysis of mechanisms for mobility, care must be taken to recognize an overconstrained mechanism. The mechanism of Fig. 1.11a has five links and six lower (revolute) pairs which indicates $F = 3(5 - 1) - 2(6) = 0$, or the mechanism has no mobility. However, since links 2, 3, and 4 are parallel to one another, removing one of the links as in Fig. 1.11b will not alter the motion of the mechanism if it should move. Then there are only four links and four lower pairs which yield $F = 3(4 - 1) - 2(4) = 1$. Figure 1.11a represents an overconstrained mechanism.

1.8 TRANSMISSION ANGLE

The transmission angle of a mechanism provides a very good indication of the quality of the mechanism, the accuracy of its performance, expected noise output, and its cost in general.

In Fig. 1.12 the angle τ is defined as the angle between the direction of the velocity of point B_2 relative to point A_2 and the direction of the

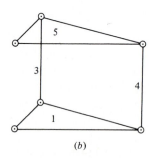

(a)   *(b)*

FIGURE 1.11
Overconstraint.

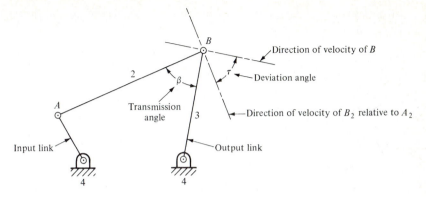

FIGURE 1.12
Transmission angle.

velocity of point B_3. The angle τ is generally referred to as the *deviation* angle. The angle β which is equal to the angle τ is more convenient to use and is called the *transmission* angle. With link 3 as the output link of the mechanism, the transmission angle influence on performance can be recognized. If the transmission angle β is equal to 90°, the force transmitted from link 2 to link 3 results in maximum torque on the output of link 3. If the transmission angle β is 0°, no torque can be realized on the output of link 3 (this is recognized as a dead center position). Various designers have differing opinions as to the proper transmission angle limits. In general, the transmission angle β should not be less than 40° nor greater than 120° for smooth operation of the mechanism. There will be exceptions to this general rule depending upon the speed of operation and the magnitude of forces transmitted. As a general rule, very small or very large transmission angles will result in large errors of motion, high sensitivity to manufacturing errors, and generally noisy and unacceptable mechanisms.

The transmission angle may be calculated from the link lengths by using the cosine relationship. In Fig. 1.13, the link lengths are identified

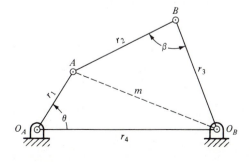

FIGURE 1.13
Transmission angle.

by a lowercase letter r with link r_1 as the input crank. The diagonal from A to O_B is of length m. Then

$$m^2 = r_4^2 + r_1^2 - 2r_4r_1 \cos \theta$$

or

$$m^2 = r_2^2 + r_3^2 - 2r_2r_3 \cos \beta$$

and

$$\cos \beta = \frac{r_2^2 + r_3^2 - r_4^2 - r_1^2 + 2r_1r_4 \cos \theta}{2r_2r_3}$$

1.9   VECTOR MANIPULATION

A vector is a quantity which has both magnitude and direction. As such it is very convenient for representation of links, displacements, velocities, accelerations, and forces. Vectors may be manipulated either graphically or analytically. The graphical representation allows for a very good representation of events and provides a nice graphical picture of the event. The analytic representation provides for rapid calculations and allows computer use. However, the analytic representation generally does not provide a good picture of the events until a final solution is reached and a plot of the results can be made. Unfortunately, the analytic representation often results in a trial-and-error solution of a synthesis problem.

1.10   GRAPHICAL VECTOR MANIPULATION

Throughout this text, vector quantities are shown as boldfaced uppercase letters. A vector is shown graphically in Fig. 1.14. It is identified as vector **A** with magnitude x indicated by the length of the vector to some convenient scale. The origin of the vector is known as the vector *tail*, while the terminus of the vector is known as the vector *head*. In general, a vector may be moved along its length, known as its line of action, without changing its magnitude or direction.

Graphical addition of vectors is accomplished by placing the tail of one vector at the head of another. Thus, in Fig. 1.15, the sum of vectors **A** and **B** is another vector, vector **C**. Note that the line of action of vector

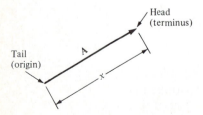

Head
(terminus)

Tail
(origin)

FIGURE 1.14
Vector representation.

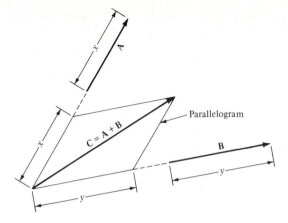

FIGURE 1.15
Graphical vector addition.

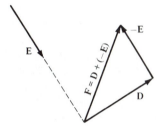

FIGURE 1.16
Graphical vector subtraction.

C must pass through the intersection of the lines of action of vectors **A** and **B**.

Graphical subtraction of vectors is accomplished by interchanging the head and tail of the vector to be subtracted and then proceeding as with addition of vectors. Thus, in Fig. 1.16, vector **D** minus vector **E** is equal to vector **F**. Note that once again the line of action of vector **F** must pass through the point of intersection of the lines of action of vectors **D** and **E**.

A series of vectors may be added or subtracted as shown in Fig. 1.17.

1.11   VECTOR MANIPULATION WITH COMPLEX NOTATION

Complex vector notation refers to representation of a vector in the imaginary plane (Argand diagram) as shown in Fig. 1.18. The vector magnitude is found by the square root of the sum of the squares of the real and imaginary components of the vector. The direction of the vector is the arctangent of the imaginary component divided by the real compo-

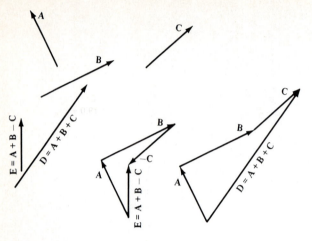

FIGURE 1.17
Vector addition and subtraction.

nent of the vector. Note that in the study of kinematics, angles are measured positively in the counterclockwise (CCW) sense.

Addition of vectors using cartesian notation is accomplished by simply adding the real parts and the imaginary parts of the vectors separately as shown in Fig. 1.19.

Subtraction of vectors using cartesian notation is accomplished by simply reversing the sign of the vector to be subtracted and proceeding as with addition of vectors as shown in Fig. 1.19.

Multiplication and division of vectors using cartesian notation is seldom encountered in analysis of mechanisms. However, the techniques for multiplication and division are included in App. A.

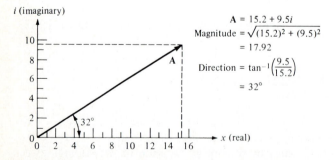

$$A = 15.2 + 9.5i$$

$$\text{Magnitude} = \sqrt{(15.2)^2 + (9.5)^2}$$

$$= 17.92$$

$$\text{Direction} = \tan^{-1}\left(\frac{9.5}{15.2}\right)$$

$$= 32°$$

FIGURE 1.18
Cartesian vector notation.

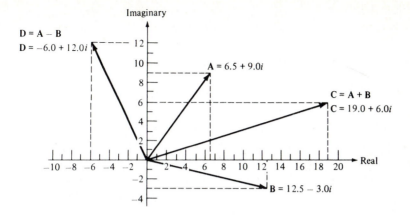

FIGURE 1.19
Adding and subtracting cartesian vectors.

1.12 VECTOR MANIPULATION WITH POLAR COMPLEX NOTATION

Polar complex notation is not very convenient for analysis of the simple addition and/or subtraction of vectors. It finds its greatest versatility in the synthesis of mechanisms by analytic means since polar vector notation allows separation of real and imaginary components of vector notation and provides two equations in place of one, thus enhancing solution of vector equations.

In Fig. 1.20, the vector **A** is represented as $Ae^{i\theta}$ in which A is the magnitude of the vector and θ is the CCW angle (measured from the x direction) indicating direction of the vector. If the real and imaginary components of the vector are considered separately, the following very important relationship can be seen:

$$Ae^{i\theta} = A(\cos \theta + i \sin \theta)$$

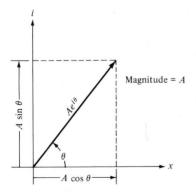

FIGURE 1.20
Polar vector notation.

In order to add (or subtract) vectors using the polar vector notation, it is necessary to convert each vector using the cos and sin notation and then proceed as with cartesian vector notation. Then the sum of two vectors **A** and **B** may be found as follows:

$$Ae^{i\theta} + Be^{i\alpha} = Ce^{i\beta}$$

$$Ae^{i\theta} = A \cos \theta + iA \sin \theta$$

$$Be^{i\alpha} = B \cos \alpha + iB \sin \alpha$$

$$Ce^{i\beta} = (A \cos \theta + B \cos \alpha) + i(A \sin \theta + B \sin \alpha)$$

The magnitude of **C** is found by the square root of the sum of the squares of the real and imaginary components of **C**. The direction of **C** is found by the arctangent of the imaginary component divided by the real component of the vector.

$$\tan \beta = \frac{A \sin \theta + B \sin \alpha}{A \cos \theta + B \cos \alpha}$$

Multiplication of a scalar quantity and a vector quantity results in a vector whose magnitude is equal to the product of the scalar and the vector magnitude and whose direction is the same as the original vector. Thus,

$$a\mathbf{A} = aAe^{i\theta}$$

where a = scalar quantity
$\mathbf{A}$ = vector quantity
aA = product of scalar and magnitude of vector

Two other types of vector multiplication are the dot product and the cross product. The dot product is indicated by placing a dot between the two vector notations as **A · B**. The dot product of two vectors results in a scalar value equal to the product of the two vector magnitudes multiplied by the cosine of the angle between their positive directions. Thus,

$$\mathbf{A} \cdot \mathbf{B} = AB \cos \theta$$

where $\mathbf{A} = Ae^{i\alpha}$ (with α larger than β)
$\mathbf{B} = Be^{i\beta}$ (with β less than α)
$\theta = \alpha - \beta$

It should be recognized that if $\mathbf{A} \cdot \mathbf{B} = 0$, then either A or $B = 0$ or the vectors are perpendicular to each other and $\cos \theta = 0$.

The dot product is distributive over addition and is commutative or

$$\mathbf{A} \cdot (\mathbf{B} + \mathbf{C}) = \mathbf{A} \cdot \mathbf{B} + \mathbf{A} \cdot \mathbf{C}$$

$$\mathbf{A} \cdot \mathbf{B} = \mathbf{B} \cdot \mathbf{A}$$

The cross product of two vectors is indicated by placing an $\times$ between the two vectors as $\mathbf{A} \times \mathbf{B}$. The result of a cross product of two vectors is another vector with magnitude equal to the product of the two vector magnitudes and the sine of the angle between the vectors. The resultant vector is perpendicular to the plane of the two vectors and sensed such that a right-hand screw turned from $\mathbf{A}$ toward $\mathbf{B}$ would advance in the direction of the resultant vector. Thus in Fig. 1.21,

$$\mathbf{A} \times \mathbf{B} = AB \sin \theta$$

where $\mathbf{A} = Ae^{i\alpha}$
$\qquad \mathbf{B} = Be^{i\beta}$
$\mathbf{A} \times \mathbf{B} = AB \sin(\alpha - \beta)$

If $\mathbf{A} \times \mathbf{B} = 0$, then either A or $B = 0$ or the two vectors $\mathbf{A}$ and $\mathbf{B}$ are parallel to each other.

The cross product is distributive over addition but is not commutative. Thus:

$$\mathbf{A} \times (\mathbf{B} + \mathbf{C}) = \mathbf{A} \times \mathbf{B} + \mathbf{A} \times \mathbf{C}$$

$$\mathbf{A} \times \mathbf{B} = -\mathbf{B} \times \mathbf{A}$$

It is sometimes convenient to refer vector expressions to a cartesian reference frame. For this purpose, vectors $\mathbf{i}$, $\mathbf{j}$, and $\mathbf{k}$ are defined as unit vectors in the x, y, and z directions, respectively. With this representation, $x\mathbf{i}$, $y\mathbf{j}$, and $z\mathbf{k}$ represent vectors with lengths x, y, and z, respectively, along the cartesian axes. A vector joining the axis origin with a point $P(x, y, z)$ can be written as (Fig. 1.22)

$$\mathbf{A} = x\mathbf{i} + y\mathbf{j} + z\mathbf{k}$$

or for any vector whose components along the x, y, and z axes are a_1, a_2, and a_3, respectively,

$$\mathbf{A} = a_1\mathbf{i} + a_2\mathbf{j} + a_3\mathbf{k}$$

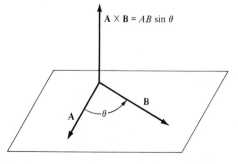

FIGURE 1.21
Vector cross product.

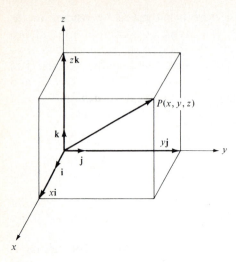

FIGURE 1.22
Vector location of a point.

Since the dot product of perpendicular vectors is always equal to 0 and the dot product of parallel unit vectors is always equal to 1, it follows that

$$\mathbf{i} \cdot \mathbf{j} = \mathbf{j} \cdot \mathbf{k} = \mathbf{i} \cdot \mathbf{k} = 0$$

and

$$\mathbf{i} \cdot \mathbf{i} = \mathbf{j} \cdot \mathbf{j} = \mathbf{k} \cdot \mathbf{k} = 1$$

Then by using the fact that dot multiplication is distributive over addition

$$\mathbf{A} \cdot \mathbf{B} = (a_1\mathbf{i} + a_2\mathbf{j} + a_3\mathbf{k}) \cdot (b_1\mathbf{i} + b_2\mathbf{j} + b_3\mathbf{k})$$

becomes

$$\mathbf{A} \cdot \mathbf{B} = a_1 b_1 + a_2 b_2 + a_3 b_3$$

If $\mathbf{A} \cdot \mathbf{B}$ is written as $AB \cos \theta$ and solved for $\cos \theta$, a useful relation results.

$$\cos \theta = \frac{a_1 b_1 + a_2 b_2 + a_3 b_3}{\sqrt{a_1^2 + a_2^2 + a_3^2} \sqrt{b_1^2 + b_2^2 + b_3^2}}$$

For the cross product of unit vectors $\mathbf{i}$, $\mathbf{j}$, and $\mathbf{k}$,

$$\mathbf{i} \times \mathbf{i} = \mathbf{j} \times \mathbf{j} = \mathbf{k} \times \mathbf{k} = 0$$

$$\mathbf{i} \times \mathbf{j} = -\mathbf{j} \times \mathbf{i} = \mathbf{k}$$

$$\mathbf{j} \times \mathbf{k} = -\mathbf{k} \times \mathbf{j} = \mathbf{i}$$

$$\mathbf{k} \times \mathbf{i} = -\mathbf{i} \times \mathbf{k} = \mathbf{k}$$

Using the fact that cross multiplication is distributive over addition, $\mathbf{A} \times \mathbf{B}$ becomes

$$\mathbf{A} \times \mathbf{B} = (a_1\mathbf{i} + a_2\mathbf{j} + a_3\mathbf{k}) \times (b_1\mathbf{i} + b_2\mathbf{j} + b_3\mathbf{k})$$

or

$$\mathbf{A} \times \mathbf{B} = (a_2b_3 - a_3b_2)\mathbf{i} - (a_1b_3 - a_3b_1)\mathbf{j} + (a_ib_2 - a_2b_1)\mathbf{k}$$

which is the expanded form of the determinant

$$\mathbf{A} \times \mathbf{B} = \begin{vmatrix} \mathbf{i} & \mathbf{j} & \mathbf{k} \\ a_1 & a_2 & a_3 \\ b_1 & b_2 & b_3 \end{vmatrix}$$

PROBLEMS

1.1. Determine the degree of freedom of each of the following mechanisms. If the degree of freedom is not 1, make recommendations for changing the mechanism.

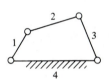

(a) General 4-bar

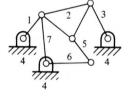

(b) Dwell mechanism

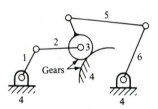

(c) Geared six-bar

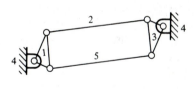

(d) Double rocker

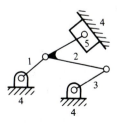

(e) Guided slider

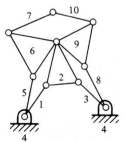

(f) Ten-bar cognate

FIGURE P1.1
Degrees of freedom.

1.2. Determine the degree of freedom of each of the following mechanisms. If the degree of freedom is not 1, make recommendations for changing the mechanism.

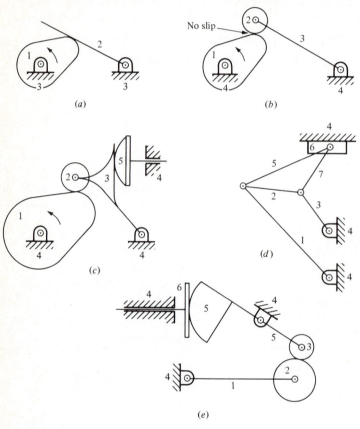

FIGURE P1.2
Degrees of freedom.

1.3. Investigate mobility of the pantograph mechanism, and explain why the mechanism will perform in a predictable manner.

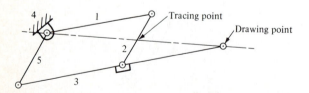

FIGURE P1.3
Mobility.

1.4. Write an interactive computer program to provide the transmission angle as a function of the input crank angle for a general four-bar mechanism. Use the program to check graphical solutions.

1.5. Graphically determine the maximum and minimum values of the transmission angle. Use the program developed in Prob. 1.4 to check the results.

$$O_AO_B = 3.0 \qquad AB = 3.5 \qquad O_AA = 1.0 \qquad O_BB = 2.0$$

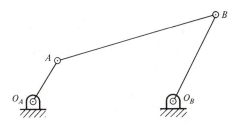

FIGURE P1.5
Transmission angle.

1.6. Graphically determine the maximum and minimum transmission angles with link 1 the input and link 5 the output.

$$O_AA = 1.0 \qquad AB = 3.15 \qquad O_BB = 1.5 \qquad BC = 1.2$$
$$AC = 2.5 \qquad CD = 2.5 \qquad O_DD = 1.5$$

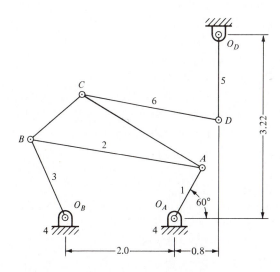

FIGURE P1.6
Limits of transmission angle.

1.7. By graphical and analytical means determine the magnitude and direction of
    (*a*) **E = A + B + C + D**
    (*b*) **E = A − B + C − D**
    (*c*) **E = B − A − C − D**
    (*d*) **E = A + B − C − D**

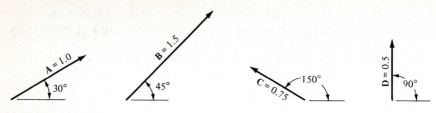

FIGURE P1.7
Vector manipulation.

1.8. Replace the forces **A**, **B**, and **C** with an equivalent force **D** showing magnitude, direction, and line of action.

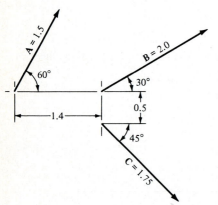

FIGURE P1.8
Vector resolution.

1.9. Determine the magnitude and direction of vector **F**.

$$\mathbf{A} = 3 + 4\mathbf{i} \qquad \mathbf{C} = -6 + 4\mathbf{i}$$
$$\mathbf{B} = 9 - 6\mathbf{i} \qquad \mathbf{D} = -8 - 6\mathbf{i}$$

(*a*) **F** = **A** + **B**   (*d*) **F** = **C** − **A**
(*b*) **F** = **B** − **C**   (*e*) **F** = **B** − **D**
(*c*) **F** = **C** − **D**   (*f*) **F** = **A** + **B** + **C** + **D**

1.10. Convert the following vectors to polar vector notation.

$$\mathbf{A} = 9 + 6\mathbf{i} \qquad \mathbf{C} = -15 + 9\mathbf{i}$$
$$\mathbf{B} = 13 - 7\mathbf{i} \qquad \mathbf{D} = -12 - 9\mathbf{i}$$

1.11. Express the vector **F** in polar vector notation.

$$\mathbf{A} = 12e^{i(75)} \qquad \mathbf{C} = 15e^{i(230)}$$
$$\mathbf{B} = 9e^{i(160)} \qquad \mathbf{D} = 6e^{i(305)}$$

(*a*) **F** = **A** + **B**   (*d*) **F** = **D** − **A**
(*b*) **F** = **B** + **C**   (*e*) **F** = **A** + **B** − **C**
(*c*) **F** = **C** + **D**   (*f*) **F** = **A** − **B** + **C** − **D**

CHAPTER
2

DISPLACEMENT
ANALYSIS

2.1 INTRODUCTION

In kinematics, displacement analysis refers to determination of the positions occupied by any or all points on a link of a mechanism as that mechanism moves through its cycle of operation. Such analysis is necessary to determine angular positions of each link for later use in force and acceleration analysis or to trace the path of a point on the coupler link. Coupler point paths are frequently used to provide desired motion of the finished machine.

A mechanism is said to have completed a *cycle* of operation when after moving through all its possible positions, it returns to its original configuration. Any position less than a complete cycle is referred to as a *phase*.

Displacement analysis assumes that the lengths of all links are known and is made easier by the original assumption that all kinematic links are rigid (except for belts and/or chains which can transmit force in one direction only). With the assumption of rigidity it is possible to locate any point on a mechanism as the mechanism operates. The analysis may be carried out through pure graphical means, algebraic means, or with the use of cartesian or polar vector notation.

2.2   GRAPHICAL DISPLACEMENT ANALYSIS

A general four-bar mechanism is shown in Fig. 2.1a. With link r_1 as the driver, the mechanism moves such that point A moves from position A_1 to position A_2 (Fig. 2.1b). It is necessary to locate the new positions of B and C. Since point B can only move in a circular arc about center O_B, an arc of radius equal to O_BB is drawn with O_B as its center. This arc indicates the locus of point B as the mechanism moves. The distance AB remains constant as the mechanism moves. Therefore, an arc of radius AB is drawn with the point A_2 as a center to intersect the arc representing the position of point B. The intersection is the new location of point B, or B_2.

In like manner, an arc of radius AC is drawn with point A_2 as its center to intersect an arc of radius BC drawn from point B_2 to locate point C_2. Thus, using only the thought that the length of links is fixed, the location of all points of interest on any mechanism as it progresses through its cycle may be determined.

Example 2.1.   Determine the path of point C in Fig. 2.2 as link r_1 completes a cycle of operation.

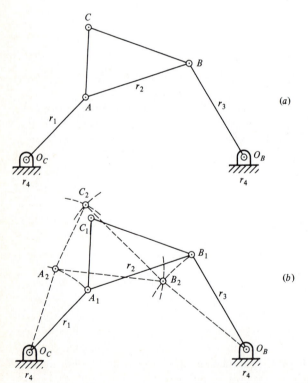

(a)

(b)

FIGURE 2.1
Link displacement.

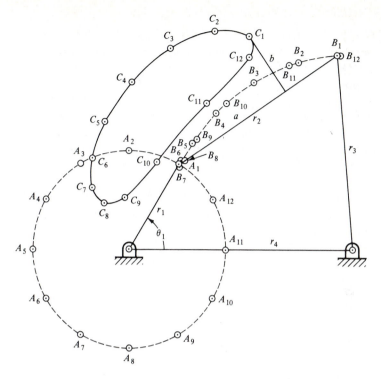

FIGURE 2.2
Example 2.1.

2.3   ANALYTIC POSITION ANALYSIS

The well-known cosine law is very useful for position analysis by analytic means. In Fig. 2.3, the cosine law provides

$$d^2 = r_4^2 + r_1^2 - 2r_1 r_4 \cos \theta_1 \qquad (2.1)$$

Considering the triangle ABO_B of Fig. 2.3,

$$d^2 = r_2^2 + r_3^2 - 2r_2 r_3 \cos \beta$$

the various angles can be found as a function of the input crank angle θ_1 as follows:

$$\cos \beta = \frac{r_2^2 + r_3^2 - d^2}{2r_2 r_3}$$

$$\cos \delta = \frac{d^2 + r_2^2 - r_3^2}{2d r_2}$$

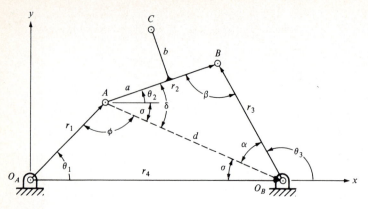

FIGURE 2.3
Vector representation.

$$\cos \alpha = \frac{d^2 + r_3^2 - r_2^2}{2r_3 d}$$

$$\cos \sigma = \frac{r_4^2 + d^2 - r_1^2}{2r_4 d}$$

$$\phi = 180 - \sigma - \theta_1$$

$$\theta_2 = \delta - \sigma \qquad\qquad (2.2)$$

Note that the angle δ could have a plus or minus sign. The two different values of δ represent the open or crossed configuration of the mechanism as shown in Fig. 2.4. In conducting position analysis by analytic means it is essential that great care be exercised to use proper angles and not to simply accept the angle returned by a hand-held calculator. Note that arccos $0.3 = \pm 72.54°$.

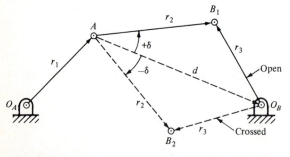

FIGURE 2.4
Branching mechanisms.

Example 2.2. Using Fig. 2.2 with $r_4 = 10$, $r_1 = 3$, $r_2 = 6$, $r_3 = 5$, $a = 4$, $b = 2$, and $\theta_1 = 57°$ determine the x and y coordinates of points A, B, and C.

Solution. From Eq. (2.1)

$$d^2 = r_4^2 + r_1^2 - 2r_1r_4 \cos \theta$$
$$= 10^2 + 3^2 - 2(10)(3) \cos 57 = 76.322$$
$$d = 8.736$$

From Eq. (2.2)

$$\cos \beta = \frac{r_2^2 + r_3^2 - d^2}{2r_2r_3}$$
$$= \frac{6^2 + 5^2 - 76.322}{2(6)(5)} = -0.2554$$
$$\beta = 104.80°$$

$$\cos \delta = \frac{d^2 + r_2^2 - r_3^2}{2dr_2}$$
$$= \frac{76.322 + 6^2 - 5^2}{2(8.736)(6)} = 0.833$$
$$\delta = 33.60°$$

$$\cos \alpha = \frac{d^2 + r_3^2 - r_2^2}{2dr_3}$$
$$= \frac{76.322 + 5^2 - 6^2}{2(8.736)(5)} = 0.748$$
$$\alpha = 41.61° \quad \text{or} \quad \alpha = 180 - \beta - \delta = 41.60°$$

$$\cos \sigma = \frac{r_4^2 + d^2 - r_1^2}{2dr_4}$$
$$= \frac{10^2 + 76.322 - 3^2}{2(10)(8.736)} = 0.958$$
$$\sigma = 16.73°$$

$$\theta_3 = 180 - \sigma - \alpha = 180 - 16.73 - 41.61 = 121.66°$$
$$\theta_2 = \delta - \sigma = 33.60 - 16.73 = 16.87°$$

Then

$$x_A = r_1 \cos \theta_1 = 3 \cos 57 = 1.634$$
$$y_A = r_1 \sin \theta_1 = 3 \sin 57 - 2.516$$
$$x_B = r_4 - r_3 \cos \phi = 10 - 5 \cos 121.66 = 12.624$$
$$y_B = r_3 \sin \phi = 5 \sin 121.66 = 4.256$$

$$x_C = x_A + a \cos \theta_2 - b \sin \theta_2$$
$$= 1.634 + 4 \cos 16.87 - 2 \sin 16.87 = 4.882$$
$$y_C = y_A + a \sin \theta_2 + b \cos \theta_2$$
$$= 2.516 + 4 \sin 16.87 + 2 \cos 16.87 = 5.591$$

2.4 POSITION ANALYSIS USING VECTOR LOOP EQUATIONS

The use of vector notation for analysis of a mechanism generally involves the cosine law although the cosine law is derived by various means. Vector notation applied to Fig. 2.5 may be used with either a calculation of the diagonal of a mechanism or by considering the complete mechanism as a closed loop of vectors as follows:

$$\mathbf{D} = \mathbf{R}_4 - \mathbf{R}_1 = \mathbf{R}_2 - \mathbf{R}_3 = de^{i\sigma} = r_4 e^{i\Omega} - r_1 e^{i\theta_1} = r_2 e^{i\theta_2} - r_3 e^{i\theta_3}$$

or $\quad \mathbf{M} = \mathbf{R}_4 + \mathbf{R}_3 = \mathbf{R}_1 + \mathbf{R}_2 = me^{i\mu} = r_4 e^{i\Omega} + r_3 e^{i\theta_3} = r_1 e^{i\theta_1} + r_2 e^{i\theta_2}$

or $\quad\quad \mathbf{0} = \mathbf{R}_1 + \mathbf{R}_2 - \mathbf{R}_3 - \mathbf{R}_4 = r_1 e^{i\theta_1} + r_2 e^{i\theta_2} - r_3 e^{i\theta_3} - r_4 e^{i\Omega}$

It is generally more convenient (although less general) to place the x axis along the fixed link and measure all angles CCW positive from the x axis. The angle Ω will then be zero.

Summing the vectors of a complete mechanism loop generally results in a rather cumbersome manipulation which will provide output after much mathematical manipulation. By using the diagonal of a

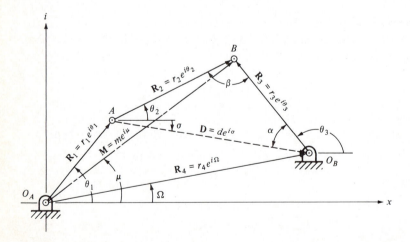

FIGURE 2.5
General vector representation.

mechanism the cosine law may be applied. This is known as Raven's method of "independent position equations" and is convenient for determination of coupler and rocker angles.

In Fig. 2.5, the diagonal from point A to O_B is designated as vector $\mathbf{D} = de^{i\sigma}$. With each link designated as a vector, it may be seen that $\mathbf{D} = \mathbf{R}_1 - \mathbf{R}_4$. Then a vector equation may be written as follows with the x axis along $O_A O_B$ and link angles measured CCW positive from the x axis.

$$r_4 = r_1 e^{i\theta_1} + de^{i\sigma} \tag{2.3}$$

Expanding using $e^{i\theta} = \cos\theta + i\sin\theta$ yields

$$r_4 = r_1(\cos\theta_1 + i\sin\theta_1) + d(\cos\sigma + i\sin\sigma)$$

Separating the real and imaginary components of the equation yields two separate equations:

$$r_{4r} = r_1\cos\theta_1 + d\cos\sigma$$
$$0 = r_1\sin\theta_1 + d\sin\sigma$$

which leads to

$$\cos\sigma = \frac{r_4 - r_1\cos\theta_1}{d} \tag{2.4}$$

$$\sin\sigma = \frac{-(r_1\sin\theta_1)}{d} \tag{2.5}$$

Dividing Eq. (2.5) by (2.4) gives the tangent of σ:

$$\tan\sigma = \frac{-r_1\sin\theta_1}{r_4 - r_1\cos\theta_1} \tag{2.6}$$

The angle σ is recognized as the angle that the diagonal AO_B makes with the horizontal or the argument of the vector D. It is unfortunate that the tangent of an angle must be used since when $\sigma = \pi/2$ or $3\pi/2$, then $\tan\sigma$ becomes infinity which gives some problems in writing computer programs unless this fact is noted. In addition, care must be taken to recognize the possibility of angles in different quadrants having the same trigonometric value (e.g., $\sin 30 = \sin 150 = 0.5$).

With the value of σ known, application of the cosine law to triangle ABO_B in Fig. 2.5 allows calculation of the angle θ_2.

$$r_3^2 = d^2 + r_2^2 - 2dr_2\cos(\sigma + \theta_2)$$

$$\sigma + \theta_2 = \arccos\frac{r_2^2 + d^2 - r_3^2}{2r_2 d} \tag{2.7}$$

Since σ is known from Eq. (2.6), the value of θ_2 becomes known.

The angle α in triangle ABO_B becomes $\alpha = 180 - \beta - (\sigma + \theta_2)$.

With d known, the angle β is available from Eq. (2.2). Then

$$\theta_3 = 180 - \alpha - \sigma$$

and $$\theta_3 = \beta + \theta_2 \tag{2.8}$$

Now the angle each link makes with the horizontal is known, and all components of each link are known.

Example 2.3. In Fig. 2.3 with $r_4 = 10$, $r_1 = 3$, $r_2 = 6$, $r_3 = 5$, and $\theta_1 = 57°$ determine the angles the coupler and rocker make with the horizontal.

Solution. From Eq. (2.1)

$$d^2 = r_4^2 + r_1^2 - 2r_1 r_4 \cos\theta_1$$
$$= 10^2 + 3^2 - 2(10)(3)\cos 57 = 76.322$$
$$d = 8.736$$

From Eq. (2.6)

$$\tan\sigma = \frac{-r_1 \sin\theta_1}{r_4 - r_1 \cos\theta_1}$$

$$= \frac{-3\sin 57}{10 - 3\cos 57} = -0.301$$

$$\sigma = \pm 16.73°$$

The plus sign indicates an open mechanism, while the minus sign indicates a crossed mechanism as shown in Fig. 2.4. In this case the plus sign is desired.
From Eq. (2.7)

$$\sigma + \theta_2 = \arccos\frac{r_2^2 + d^2 - r_3^2}{2r_2 d}$$

$$= \arccos\frac{6^2 + 76.322 - 5^2}{2(6)(8.736)}$$

$$= 33.59°$$

and $$\theta_2 = 33.59 - 16.73 = 16.86°$$

From Eq. (2.2)

$$\cos\beta = \frac{r_2^2 + r_3^2 - d^2}{2r_2 r_3}$$

$$= \frac{6^2 + 5^2 - 76.322}{2(6)(5)} = -0.2554$$

$$\beta = 104.80°$$

and $$\theta_3 = \beta + \theta_2$$

$$= 104.88 + 16.86 = 121.66°$$

2.5 BASIC PROGRAM FOR CALCULATING LINK ANGLES

The angles links of a four-bar mechanism make with the horizontal may be calculated with the use of an interactive computer program (ANGLES, App. A).

Output from the program will be

* * * * * * * * * * * * * * * * * * *   ANGLES   * * * * * * * * * * * * * * * * * * *

THIS PROGRAM PROVIDES ANGLES WITH THE HORIZONTAL
(COUNTERCLOCKWISE POSITIVE) FOR EACH LINK OF AN OPEN OR
CROSSED MECHANISM. ALL ANGLES ARE IN DEGREES.

LENGTH OF FIXED LINK = 10
LENGTH OF CRANK   = 4
LENGTH OF COUPLER   = 12
LENGTH OF ROCKER   = 8

MECHANISM IS OPEN

CRANK ANGLE	COUPLER ANGLE	ROCKER ANGLE	TRANSMISSION ANGLE
0	36	62	26
30	22	55	32
60	18	64	46
90	18	80	61
120	21	96	74
150	27	110	83
180	34	121	86
210	44	127	83
240	54	128	74
270	62	123	61
300	65	111	46
330	56	89	32
360	36	62	26

2.6 COUPLER CURVES

A path traced by a point on the coupler link as the mechanism completes one full cycle is known as a *coupler* curve.

The *Analysis of Four-Bar Linkages* by Hrones and Nelson presents an excellent picture of coupler curves associated with crank rocker mechanisms. The atlas includes over 7000 coupler curves and is particularly useful for designers who are searching for a particular shape of coupler curve. The coupler curves are drawn in dashed lines with each dash length representing a 5° rotation of the input crank. Thus short dashes represent low velocity and longer dashes represent higher velocity of the coupler point. With this feature, a designer receives a large-scale representation of the coupler curve shape, the length of the links which generate the coupler curves, and an idea of the velocity of the coupler point along with some visual indication of the acceleration associated with the coupler point.

A general equation of a coupler curve may be developed by reference to Fig. 2.6. Point C with the coordinates (x, y) is the point on the coupler which will trace a coupler curve as the input crank r_1 rotates. Then

$$
\begin{aligned}
x = x_A + a \cos \Gamma \qquad &\text{or} \qquad x_A = x - a \cos \Gamma \\
y = y_A + a \sin \Gamma \qquad &\text{or} \qquad y_A = y - a \sin \Gamma
\end{aligned}
\tag{2.9}
$$

and

$$
\begin{aligned}
x = x_B + b \cos (\Gamma + \mu) \qquad &\text{or} \qquad x_B = x - b \cos (\Gamma + \mu) \\
y = y_B + b \sin (\Gamma + \mu) \qquad &\text{or} \qquad y_B = y - b \sin (\Gamma + \mu)
\end{aligned}
\tag{2.10}
$$

Since points A and B are restricted to move in circles with radii q and s, respectively,

$$
\begin{aligned}
r_1^2 &= (x - a \cos \Gamma)^2 + (y - a \sin \Gamma)^2 \\
r_3^2 &= [x - b \cos (\Gamma + \mu) - p]^2 + [y - b \sin (\Gamma + \mu)]^2
\end{aligned}
\tag{2.11}
$$

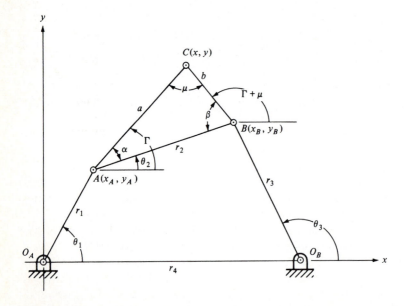

FIGURE 2.6
Coupler curve development.

Expanding Eqs. (2.11) and using trigonometric identities will yield $\sin \Gamma$ and $\cos \Gamma$ which may be substituted into the identity $\sin^2 \Gamma + \cos^2 \Gamma = 1.0$. This will result (after very much algebraic manipulation) into a formidable sixth-order equation in x and y which is the general equation of a coupler curve. Hartenberg and Denavit[*] have reported the general equation to be

$$
\begin{aligned}
&\{\sin \alpha [(x - r_4) \sin \mu - y \cos \mu](x^2 + y^2 + a^2 - r_4^2) \\
&\quad + (y \sin \beta)[(x - r_4)^2 + y^2 + b^2 - r_3^2]\}^2 \\
&+ \{\sin \alpha [(x - r_4) \cos \mu + y \sin \mu](x^2 + y^2 + a^2 - r_1^2) \\
&\quad - (x \sin \beta)[(x - r_4)^2 + y^2 + b^2 - r_3^2]\}^2 \\
&= 4b^2(\sin \alpha)(\sin^2 \beta)(\sin^2 \mu)[x(x - r_4) - y - r_4 y \cot \mu]^2
\end{aligned}
$$

The general equation for a coupler curve is very unwieldy; however, it is useful for investigating coupler curves of a particular nature.

A significant deduction is that, if a coupler curve contains double points (points at which two or more tangents to the coupler curve appear), those double points will all lie on a circle with the equation:

$$
x^2 + y^2 - r_4 x - r_4 y \cot \mu = 0
$$

This circle is known as the circle of singular foci and passes through the fixed pivots O_A and O_B. An interesting feature of double points is that they will all see the fixed link of a mechanism through the same angle as shown in Fig. 2.7.

2.7 SYMMETRIC COUPLER CURVES

The path of a coupler point is often used as input motion for a second mechanism or is used as the path of an insertion mechanism in an automated machine. For this reason it may be advantageous to generate a coupler curve which is symmetric about some axis. If the coupler AB of a mechanism is made the same length as the output rocker $O_B B$ and the coupler point C is made the same distance from point B, the path of the coupler point will be symmetric as shown in Fig. 2.8.

The axis of symmetry is the line through the coupler point and the output crank fixed pivot when the input crank is in alignment with the two fixed pivots. In Fig. 2.8 with A_1 and A_2 symmetric with A which is in alignment with the fixed centers, a circle of radius AB may be drawn from B_1 and will pass through points C_1, A_1, and center O_B. A circle of the

---

[*] R. S. Hartenberg and J. Denavit, *Kinematic Synthesis of Linkages*, McGraw-Hill, 1964, p. 151.

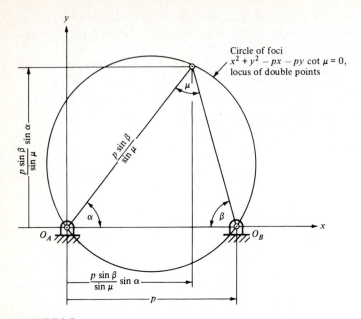

FIGURE 2.7
The circle of singular foci.

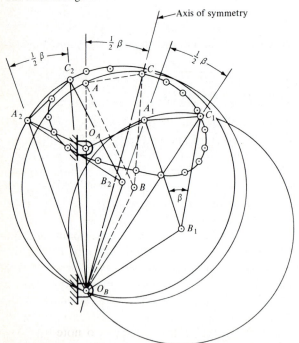

FIGURE 2.8
Symmetric coupler curve.

same radius centered at B_2 will pass through A_2, C_2, and O_B. The angle $A_1B_1C_1$ will be equal to the angle $A_2B_2C_2$ which is called β. Then the angle $A_1O_BC_1$ will be equal to the angle AO_BC and to the angle $A_2O_BC_2$ or will be equal to $\frac{1}{2}\beta$. Thus the axis of symmetry must be the line O_BC in Fig. 2.8.

Symmetric coupler curves are often convenient for use in automated assembly systems or in driving indexing systems.

2.8   BASIC PROGRAM FOR GENERATION OF COUPLER CURVES

Coupler curves may be computer generated using a few additions to the program ANGLES to create the program COUPLER (App. A).

Output from the modified program will be

LENGTH OF FIXED LINK = 10
LENGTH OF CRANK   = 4
LENGTH OF COUPLER   = 12
LENGTH OF ROCKER   = 8
DISTANCE ALONG COUPLER TO PERPENDICULAR TO COUPLER POINT 6
PERPENDICULAR DISTANCE FROM COUPLER TO COUPLER POINT   4

MECHANISM IS OPEN

CRANK ANGLE	COUPLER ANGLE	ROCKER ANGLE	TRANSMISSION ANGLE	COUPLER	
				CX	CY
0	36	62	26	5.618019	1.175592
30	22	55	32	5.318464	2.749231
60	18	64	46	3.902123	4.082153
90	18	80	61	1.902138	4.618063
120	21	96	74	−0.1328108	4.180888
150	27	110	83	−1.682073	2.908055
180	34	121	86	−2.341940	1.118476
210	44	127	83	−2.025479	−0.6105971
240	54	128	74	−0.8245263	−1.846009
270	62	123	61	0.9388232	−2.234095
300	65	111	46	2.845122	−1.651533
330	56	89	32	4.582407	−0.3420184
360	36	62	26	5.618019	1.175592

2.9   COGNATE MECHANISMS

If a coupler curve is desired and a mechanism is synthesized by some means to produce that coupler curve, it is interesting to note that two other mechanisms which will produce the same coupler curve exist. These two mechanisms are known as *cognate linkages*. In some cases a mechanism will be devised to accomplish a given task but cannot be used since the fixed pivot locations are not suitable for some reason. Then it is entirely possible that one of the cognate mechanisms can be used.

Figure 2.9 shows a general four-bar mechanism $O_A A B O_B$ with coupler point C. The mechanism has been designed such that point C will trace a desired path. The line $O_A D$ is made parallel to line AC. The line CD is made parallel to line $O_A A$. Triangle DCF is made similar to coupler triangle ABC. The line $O_B E$ is made parallel to the line BC. The line CE is made parallel to the line $O_B B$. Triangle ECG is made similar to the coupler triangle ABC. Line FH is made parallel to line CG, and line GH is made parallel to line CF. This construction is identified by the triangles of Fig. 2.10. In Fig. 2.10 the line FB is parallel to the line HO_B through point C. The line GA is parallel to the line HO_A through point C. The line DE is made parallel to the line $O_A OB$ through point C. Equal angles are easily recognized.

Figure 2.9 identifies three mechanisms which will trace the same coupler curve with point C as the coupler point. The cognate mechanisms are $O_A DFH$ and $O_B EGH$. Proof of the construction involves computation of the coordinates of point H as demonstrated by Roberts.[*]

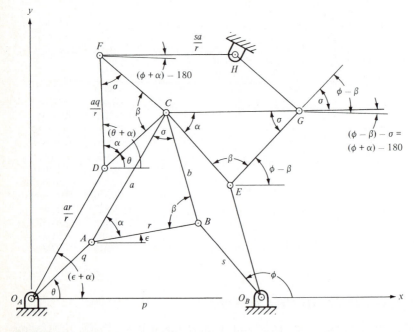

FIGURE 2.9
Cognate mechanisms.

---

[*] S. Roberts, On Three-Bar Motion in Plane Space, *Proc. London Math. Soc.*, vol. 7, 1876.

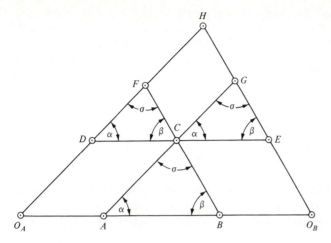

FIGURE 2.10
Cognate construction guide.

The length $O_A D$ in Fig. 2.9 is ar/r and makes the angle $\epsilon + \alpha$ with the horizontal. The length DF is aq/r and makes the angle $\theta + \alpha$ with the horizontal. The length FH is sa/r and makes the angle $\phi + \alpha - 180$ with the horizontal. Then the x coordinate of H becomes

$$x_H = \frac{ar}{r} \cos (\epsilon + \alpha) + \frac{aq}{r} \cos (\theta + \alpha) - \frac{sa}{r} \cos (\phi + \alpha)$$

$$= \frac{ar}{r} (\cos \epsilon \cos \alpha - \sin \epsilon \sin \alpha)$$

$$+ \frac{aq}{r} (\cos \theta \cos \alpha - \sin \theta \sin \alpha)$$

$$- \frac{sa}{r} (\cos \phi \cos \alpha - \sin \phi \sin \alpha)$$

$$= \frac{a}{r} (r \cos \epsilon + q \cos \theta - s \cos \phi) \cos \alpha$$

$$- \frac{a}{r} (q \sin \theta + r \sin \epsilon - s \sin \phi) \sin \alpha$$

From the basic four-bar mechanism of Fig. 2.6,

$$r \cos \epsilon + q \cos \theta - s \cos \phi = p$$

$$q \sin \theta + r \sin \epsilon - s \sin \phi = 0$$

Then

$$X_H = \frac{a}{r} p \cos \alpha$$

In a similar manner the y component of H becomes

$$y_H = \frac{a}{r}\, p \sin \alpha$$

The distance $O_A H$ becomes

$$O_A H = \sqrt{x_H^2 + y_H^2} = \frac{a}{r}\, p$$

and the angle $O_B O_A H$ becomes

$$O_B O_A H = \arctan \frac{y_H}{x_H} = \alpha$$

Now, since the x and y coordinates of the point H contain no variables, the point H is a fixed point on the original four-bar mechanism, and both cognate mechanisms pass through a complete cycle of operation.

Example 2.4. Determine the cognate mechanisms of the four-bar mechanism $O_A A B O_B$ in Fig. 2.11.

Solution
Note: Angle ABC = angle DCF = angle CEG = 180°, and angle BAC = angle CDF = angle ECG = 0°.

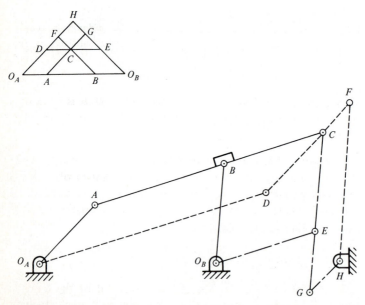

FIGURE 2.11
Example 2.4.

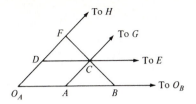

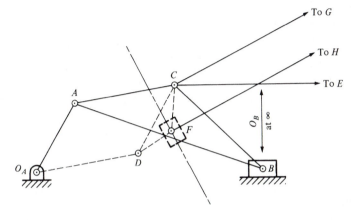

FIGURE 2.12
Example 2.5.

Example 2.5. Determine the cognate mechanism of the slider crank mechanism in Fig. 2.12. (Note that only one cognate of a slider crank mechanism exists.)

Solution
Note: Since the center of curvature of the path of point B is located at infinity in a direction perpendicular to the path of B, it is not possible to draw the complete triangle system of Fig. 2.10.

With the parallel line construction process, it may be seen that the angular velocity of link $O_A D$ of the left cognate mechanism of Fig. 2.8 will have the same angular velocity as link q of the parent mechanism. In the right cognate mechanism, link HG will have the same angular velocity as link q of the parent mechanism.

2.10 ADJUSTABLE MECHANISMS

A mechanism designed so that the length of any or all of the links is adjustable provides for either fine tuning of the completed mechanism or for producing desirable output features of the system.

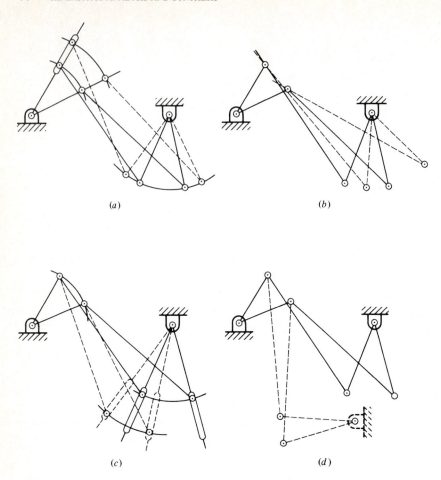

(a)   (b)

(c)   (d)

FIGURE 2.13
Adjustable links.

In the case of a dwell mechanism (Fig. 2.15), the length of dwell of the follower may be adjusted by altering the fixed pivots locations.

In Fig. 2.13a, the input crank is made of adjustable length which will have the general effect of increasing the output link angular displacement. In Fig. 2.13b, the coupler link is made of adjustable length which will have the general effect of repositioning the output crank and slightly influencing its angular displacement. In Fig. 2.13c, the output crank is made of variable length which will have the general effect of repositioning the output crank and slightly influencing its angular displacement. Figure 2.14d shows the influence of adjusting the fixed pivot location. The greatest influence of adjustable fixed pivots is in altering the coupler curve shape as in the case of an adjustable dwell mechanism (Fig. 2.14).

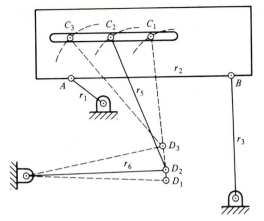

FIGURE 2.14
Adjustable dwell mechanism.

In Fig. 2.14, link 5 may be adjusted to connect to link 2 at either point C_1, C_2, or C_3. Link 5 may be adjusted so that the position of link 6 is not influenced. As link 1 rotates, link 6 will oscillate except when the coupler point C is moving on a circular arc centered at point D. Since the length of the circular arc of coupler curves C_1, C_2, and C_3 is not the same, the length of time that link 6 does not move is altered.

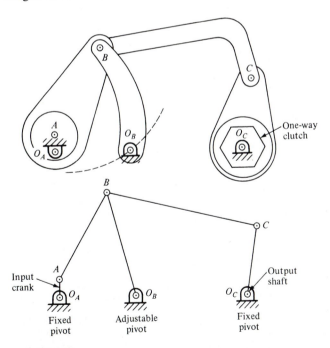

FIGURE 2.15
Zero-Max speed changer.

Repositioning the fixed pivots will have a greater influence on mechanism performance than will altering the link lengths.

The Zero-Max speed reducer shown in Fig. 2.15 is an excellent example of an adjustable mechanism in which the angular displacement of the output crank is made adjustable while the mechanism is in operation. This has the effect of providing variable angular velocity of the output shaft with constant angular velocity of the input crank shaft. Since a single mechanism would provide intermittent output rotation, a series of three or more slightly out of phase mechanisms may be operated on the same input and output shaft.

PROBLEMS

2.1. Determine the path of point C on the coupler of the Watt mechanism for $\theta_1 = \pm 45°$ from a given position.

$$O_A A = AC = O_B B = 2.0 \qquad AB = 4.0$$

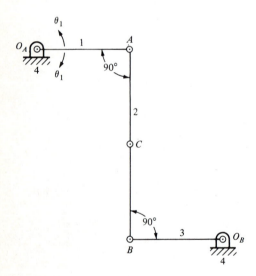

FIGURE P2.1
Watt mechanism.

2.2. Determine the path of point C using 15° increments of rotation of link 1.

$$O_A A = 1.0 \qquad AB = 3.0 \qquad O_B B = 1.5$$
$$O_A O_B = 3.0 \qquad AC = BC = 2.0$$

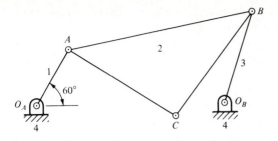

FIGURE P2.2
Coupler curve.

2.3. (a) Plot the coupler curve associated with point C as link 1 makes one complete revolution in 15° increments.

(b) Move O_A 0.5 to the left and 0.5 down and replot the coupler curve associated with point C.

$$O_AO_B = 3.0 \qquad O_AA = 1.5 \qquad AB = 3.5$$
$$O_BB = 2.0 \qquad AC = 3.5 \qquad BC = 2.0$$

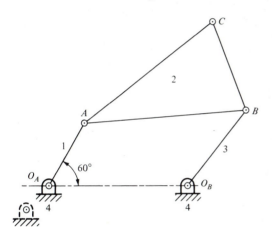

FIGURE P2.3
Coupler curves.

2.4. Plot coupler curve associated with point C as link 1 makes one complete revolution in 15° increments.

$$O_AA = 1.0 \qquad O_BB = 2.0 \qquad AB = 2.5 \qquad AC = 5.0$$

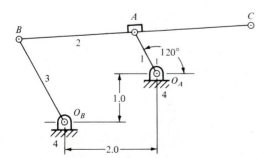

FIGURE P2.4
Coupler curve.

2.5. Derive equations for computation of x and y coordinates of point C in Prob. 2.4. Origin of the coordinate system is at O_A.

2.6. Develop equations for the angles θ_2, θ_3, and β in terms of angle θ_1 and link lengths.

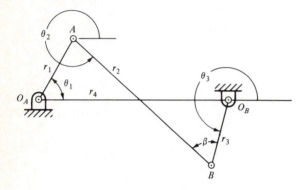

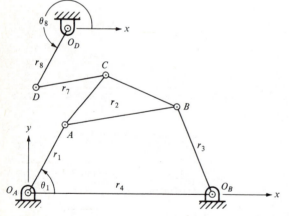

FIGURE P2.6
Angular displacements.

2.7. Derive equations for plotting the angle θ_8 as a function of θ_1. Location of point O_D is known.

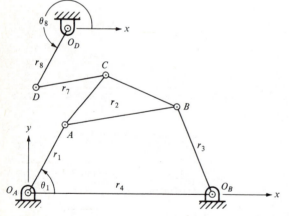

FIGURE P2.7
Angular displacement.

2.8. Derive equations for displacement of the piston (link 6) as a function of angle θ_1.

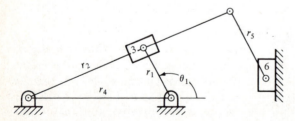

FIGURE P2.8
Linear diplacement.

2.9. Find the locus of point C and its axis of symmetry by graphical means using $15°$ increments of θ_1.

$$O_AA = 0.75 \qquad AB = 2.00 \qquad AC = 3.75$$
$$O_BB = 2.00 \qquad BC = 2.00$$

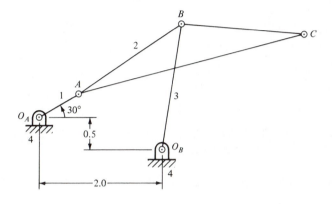

FIGURE P2.9
Symmetric coupler curve.

2.10. Find the locus of point C and its axis of symmetry using $15°$ increments of θ_1.

$$O_AO_B = 2.10 \qquad O_AA = 0.75 \qquad O_BB = 1.55$$
$$AB = 1.55 \qquad BC = 1.55 \qquad AC = 3.00$$

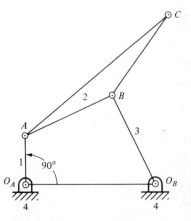

FIGURE P2.10
Symmetric coupler curve.

2.11. Locate the cognates of the following mechanisms.

(a) $O_A O_B = 13.0$ $O_A A = 6.0$ $O_B B = 9.0$

       $AB = 10.0$ $AC = 4.5$ $BC = 8.5$

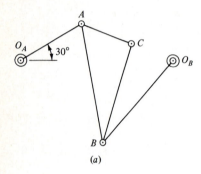

(a)

(b) $O_A O_B = 5.25$ $O_A A = 2.25$ $O_B B = 4.00$

       $AB = 4.75$ $AC = 2.25$

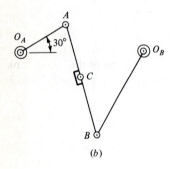

(b)

(c) $O_A O_B = 5.25$ $O_A A = 2.00$ $O_B B = 2.00$

       $AC = 2.35$ $BC = 2.35$

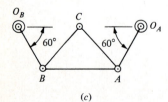

(c)

FIGURE P2.11
Cognates.

2.12. The mechanism shown has been designed for use in an automated assembly unit. An insertion tool is located at point C and must move in the path determined by the mechanism. Unfortunately, the bearing located at point O_A interferes with another portion of the unit. Relocate the bearing O_A without altering the path of the insertion tool. (Try using a cognate.)

$$O_A O_B = 8.0 \qquad O_A A = 3.0 \qquad O_B B = 3.0$$
$$CD = 8.0 \qquad AD = DB$$

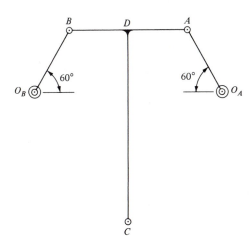

FIGURE P2.12
Insertion tool bearing location.

2.13. Using complex polar notation, derive an equation for the position of the piston in an offset slider crank mechanism as a function of the input crank angle θ_1. Explain the meaning of two values of the piston position which result from solution of a quadratic equation.

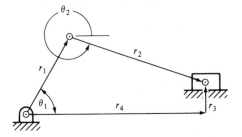

FIGURE P2.13
Slider crank piston displacement.

CHAPTER

3

VELOCITY

3.1 CONCEPTS

Velocity is defined as the time rate of change of position of a point. As such, velocity is a vector quantity which has both magnitude and direction. Speed, on the other hand, does not have direction but only magnitude. An airplane is said to have a speed of 550 miles per hour (mi/h) or a velocity of 550 mi/h in a northeast direction.

A velocity may be an *average* velocity which is computed with the time interval very large, or a velocity may be *instantaneous* which is computed with the time interval very small. In the study of kinematics, interest is centered on the instantaneous velocity.

If the velocity of a point is compared to a fixed (nonmoving) point, the velocity is said to be an *absolute* velocity. If the velocity of a point is compared to another moving point, that velocity is said to be a *relative* velocity. A passenger in an airplane who is walking forward in a plane at a velocity of 1 mi/h relative to the airplane while the plane is traveling at a velocity of 550 mi/h in a northeast direction will be traveling at a velocity of 551 mi/h in a northeast direction.

Since velocity is a vector quantity, it may be resolved into components in selected directions and may be added or subtracted following the rules of vector manipulation. *Linear* velocity refers to the velocity of a point on a mechanism. *Angular* velocity refers to the rate of rotation of a link of a mechanism.

In Fig. 3.1, the point A is located by position vector $\mathbf{A_1}$ with magnitude a_1 at time 1. At time 2 the point A is located by position vector $\mathbf{A_2}$ with magnitude a_2. The change in position of point A from position 1 to position 2 is then the vector difference $\mathbf{A_2} - \mathbf{A_1}$ or $\Delta\mathbf{A}$. In Fig. 3.1, the distance $\mathbf{OB}$ is made equal to the distance $\mathbf{OA_1}$ and the angle $\Delta\theta$ is very small. Then the angle A_1BA_2 is almost 90°, and the limit as $\Delta\theta$ approaches zero becomes equal to 90°.

The velocity of point A as it moves from position 1 to position 2 is then the limit as Δt approaches zero of $\Delta\mathbf{A}/\Delta t$ or is the limit as Δt approaches zero of $\mathbf{A_1B}/\Delta t + \mathbf{BA_2}/\Delta t$.

The limit of $\mathbf{A_1B}/\Delta t$ as Δt approaches zero is

$$\lim_{\Delta t \to 0} \frac{\mathbf{A_1B}}{\Delta t} = \lim_{\Delta t \to 0} \frac{a\,\Delta\theta}{\Delta t} = a\,\frac{d\theta}{dt}$$

$d\theta/dt$ is the angular velocity of the vector $\mathbf{A}$ measured in radians per unit time and is generally given the symbol ω. Note that the distance a represents the distance from the head of vector $\mathbf{A_1}$ to its origin or point of rotation.

The limit of $\mathbf{BA_2}/\Delta t$ as Δt approaches zero is

$$\lim_{\Delta t \to 0} \frac{\mathbf{BA_2}}{\Delta t} = \lim_{\Delta t \to 0} \frac{a_2 - a_1}{\Delta t} = \frac{da}{dt}$$

da/dt is recognized as the rate at which the position vector is changing magnitude (increasing or decreasing).

The velocity of point A in moving from position 1 to position 2 becomes the vector sum of two components. One component of the velocity is due to a change in direction of the position vector, and the other component is due to a change in magnitude of the position vector.

$$V_A = a\omega + \frac{da}{dt} \tag{3.1}$$

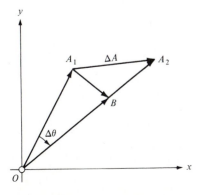

FIGURE 3.1
Displacement of a point.

As the angle $\Delta\theta$ approaches zero (or as the time interval becomes very small), the angle A_1BA_2 becomes 90° and the two vectors which make up the components of the velocity of point A become orthogonal. More important, the vector $a\omega$ becomes perpendicular to the original position vector $\mathbf{A}_1$ and is directed in the sense indicated by the changing position angle of the vector $\mathbf{A}$.

3.2 VELOCITY ANALYSIS BY THE METHOD OF COMPONENTS

Since velocity is a vector quantity, its resolution into components allows for analysis of a mechanism for linear and angular velocity of its components.

In Fig. 3.2, link 1 is rotating counterclockwise with the angular velocity ω_1. By Eq. (3.1), since the length of link 1 is constant, the linear velocity of point A becomes $O_A A\omega_1$ and is directed perpendicular to link 1 in the direction indicated by ω_1 as shown. The velocity of point A is given the symbol V_A.

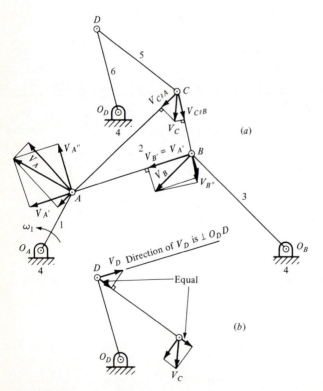

FIGURE 3.2
Velocity components.

The velocity V_A may be resolved into two components. One component is parallel to the line AB (indicated as $V_{A'}$) and the other component is perpendicular to the line AB (indicated as $V_{A''}$). The line AB represents one element of link 2 and as such is fixed in length. If point A on line AB has a velocity $V_{A'}$ parallel to its length and the length of the line cannot change, then point B must have a velocity such that its component parallel to line AB will be equal to $V_{A'}$ (indicated as $V_{B'}$). Point B is located on link 3 as well as on link 2, and by Eq. (3.1) its velocity must be in a direction perpendicular to link 3. In order for point B to have velocity component $V_{B'}$ parallel to line AB and have its direction as shown, the velocity component $V_{B''}$ perpendicular to line AB must be as shown.

The velocity of point C on link 2 may be found by using components of the velocity of points A and B as follows. Resolving the velocity V_A into two components, one parallel to the line AC and the other perpendicular to the line AC, and using the concept that the distance AC is constant, a component of the velocity C parallel to the line AC becomes $V_{C\|A}$. This is a component of the velocity of point C relative to point A and parallel to the line AC. By resolving the velocity of point B into components parallel and perpendicular to the line BC, the component of the velocity of C parallel to the line BC may be found as $V_{C\|B}$. The velocity of point C is found as that velocity which has the two components $V_{C\|A}$ and $V_{C\|B}$ as shown.

In a similar manner, the velocity of point D may be found as shown in Fig. 3.2. With the velocity of point D known, the angular velocity of link 6 can be found using Eq. (3.1):

$$\omega_6 = \frac{V_D}{O_D D}$$

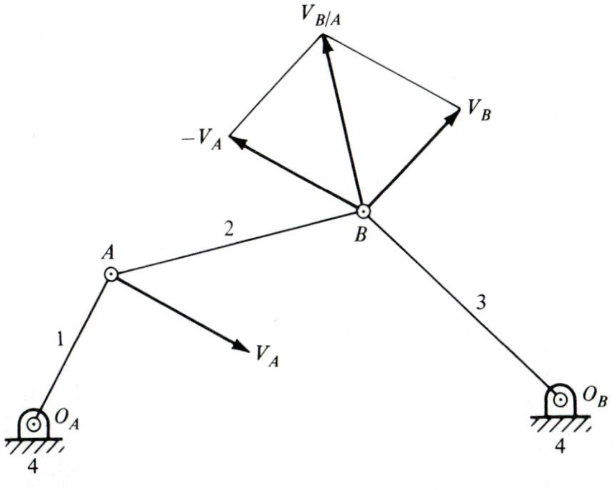

FIGURE 3.3
Relative velocity.

If, as shown in Fig. 3.3, the velocity of point A is subtracted from the velocity of every point on the mechanism, the velocity of point A becomes zero and the velocity of point B becomes directed perpendicular to the line AB and is known as the velocity of B with respect to A, or $V_{B/A}$. This is necessary since point A then becomes a fixed point, the distance AB is constant, and point B is essentially rotating about point A as a center. Then, by Eq. (3.1), the angular velocity of link 2 will become

$$\omega_2 = \frac{V_{B/A}}{Ab}$$

The angular velocity of link 5 of Fig. 3.2 may be found in a similar manner by subtracting the linear velocity of point D from that of point C.

3.3 INSTANT CENTERS FOR VELOCITY

Following the thoughts of Reuleaux*, all points on a mechanism may be considered as moving at any instant with rotation about some center. The center for rotation is the instant center for velocity. In the case of a point moving in a straight line, the center of rotation for that point may be considered as being located at infinity in a direction perpendicular to the line of motion of the point.

For a link which is rotating about a fixed center, as for instance links 1 and 3 in Fig. 3.4, the instantaneous center for velocity is located at the fixed center of rotation (O_A and O_B for links 1 and 3). Points $P_{(1,4)}$ and $P_{(3,4)}$ are instantaneous centers of rotation for links 1 and 3. Links 1 and 2 and links 2 and 3 are connected with revolute joints at points A and B. Thus point A becomes the instantaneous center for velocity between links

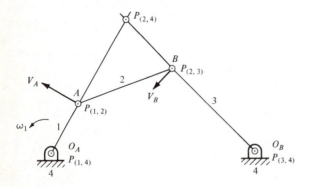

FIGURE 3.4
Instant center for velocity.

* F. Reuleaux, *The Kinematics of Machinery*, Macmillan, New York, 1876.

1 and 2, or $P_{(1, 2)}$, and point B becomes the instantaneous center for velocity between links 2 and 3, or $P_{(2, 3)}$.

The velocity of a point rotating about a fixed center on a link of fixed length must, by Eq. (3.1), be directed perpendicular to that link and in a direction indicated by the angular velocity of the link.

In Fig. 3.4 a general four-bar mechanism is shown in which link 1 is rotating with angular velocity ω_1 counterclockwise. The velocity of point A on link 2 (or on link 1 since point A is a revolute joint) must be directed perpendicular to link 1. In like manner, the velocity of point B on link 2 must be directed perpendicular to link 3. Considering now link 2 as a "free body" with the velocity of points A and B on link 2 known, perpendiculars drawn to the velocity vectors will of necessity intersect at the center about which link 2 is instantaneously rotating. Such a center as point $P_{(2, 4)}$ is known as the instant center for velocity of link 2. The center $P_{(2, 4)}$ refers to the motion of link 2 with respect to link 4 or the motion of link 4 with respect to link 2.

The instant center $P_{(2, 4)}$ is in reality two coincident points. One point is on link 2, and the other point is on link 4. These two coincident points must have the same velocity (in the case of $P_{(2, 4)}$ that velocity is zero since all points on link 4 have zero velocity).

It is important to note that point $P_{(2, 4)}$ is the instant center used for velocity analysis only. The position of the instant center for velocity is easily found by simply extending links 1 and 3 to locate their intersection.

With the instant center for velocity of link 2 known, the angular velocity of link 2 may be found using Eq. (3.1). Thus

$$\omega_2 = \frac{V_A}{AP_{(2, 4)}}$$

With the angular velocity of link 2 known, the linear velocity of any point on link 2 may be found as the product of its distance from $P_{(2, 4)}$ and the angular velocity of link 2. This seems to imply that the distance to the instant center for velocity is a constant distance. However, this is an instantaneous situation and in reality the distance to the instant center for velocity is changing.

If the mechanism of Fig. 3.4 is inverted to allow link 4 to move and to fix link 1, the mechanism of Fig. 3.5 results. Since this is a simple inversion, the relative motion of the links is unchanged. With this configuration, links 4 and 2 become links rotating about fixed centers. Their lengths must not change during any motion, and the instantaneous center for velocity of link 3 relative to link 1 is found by simply extending links 2 and 4 to their intersection. This procedure locates instant center $P_{(1, 3)}$ which relates the motion of link 3 with respect to link 1, or vice versa. It is also true that point $P_{(1, 3)}$ is in reality two coincident points, one on link 1 and one on link 3 and that these two coincident points, at this instant, have exactly the same velocity.

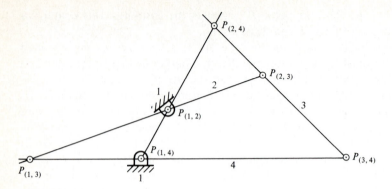

FIGURE 3.5
Inverted mechanism of Fig. 3.4.

Since there exists an instantaneous center for velocity relating the motion of each link with that of another link in a mechanism, the total number of instant centers existing becomes

$$N = \frac{n(n-1)}{2}$$

where n is the number of links in a mechanism.

A four-bar mechanism will have $4(4-1)/2 = 6$ instant centers for velocity. An eight-bar mechanism will have $8(8-1)/2 = 28$ instant centers for velocity.

Fortunately it is often not necessary to locate all the instantaneous centers for velocity to analyze a mechanism. A good designer will locate only those centers necessary to provide the desired information. By application of the Aronhold-Kennedy rule it is possible to locate all instant centers for velocity with a simple graphical construction.

Figure 3.6 indicates two links of a mechanism each of which is considered to be instantaneously rotating about its center for velocity. Link 2 is rotating about center $P_{(1,2)}$ with angular velocity ω_2 and link 6 is rotating about center $P_{(1,6)}$ with angular velocity ω_6.

The linear velocity of point A on link 2 is equal to the distance from point A to center $P_{(1,2)}$ multiplied by the angular velocity of link 2 and is shown in Fig. 3.6 as V_A. The linear velocity of point B on link 2 which is the same distance from $P_{(1,2)}$ as is point A but located on the line joining $P_{(1,2)}$ and $P_{(1,6)}$ must be equal to the linear velocity of point A and is shown as V_B. Since the velocity of any point on a rotating link is directly proportional to its distance from the center of rotation, a simple graphical triangulation technique will provide the velocity of a point on link 2 such as that of point C. In a similar manner, the velocity of points D, E, and F on link 6 may be determined.

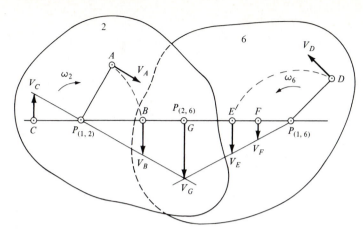

FIGURE 3.6
Aronhold-Kennedy rule.

Point G of Fig. 3.6 will have the velocity V_G if it is located on link 2 and will have the same velocity if it is located on link 6, and that velocity vector will be directed perpendicular to the line joining instantaneous centers $P_{(1, 2)}$ and $P_{(1, 6)}$. Therefore, point G represents the location of two coincident points, one on link 2 and one on link 6, which have exactly the same velocity. By definition then, point G becomes the instantaneous center for velocity $P_{(2, 6)}$. With three links represented (links 1, 2, and 6) the number of instantaneous centers for velocity becomes $3(3 - 1)/2 = 3$. The three instantaneous centers for velocity associated with three links of a mechanism must lie on a straight line. This is the Aronhold-Kennedy rule.

The fact that the three instant centers for velocity associated with three links of a mechanism lie on a straight line makes graphical location of the instant centers possible. It is necessary to remember the two definitions of the instant center for velocity:

1. The instant center for velocity consists of a point on one link which may be considered as a center of rotation of another link.
2. The instant center for velocity consists of two coincident points on two separate links which at the particular instant have exactly the same velocity.

Figure 3.7 shows a slider crank mechanism for which the location of all instant centers for velocity is desired. The slider crank is a four-bar mechanism, and there will be $4(4 - 1)/2 = 6$ instant centers for velocity.

In almost all mechanisms, many of the instant centers can be located by inspection. Thus, instant center $(1, 4)$ is the revolute connection

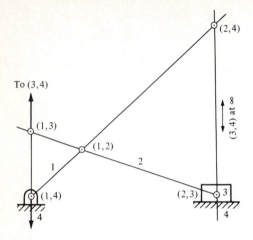

FIGURE 3.7
Instant centers for a slider crank.

between links 1 and 4 since this point is a point on link 4 about which link 1 must rotate. Since links 1 and 2 are connected with revolute pairs, the revolute pair consists of two coincident points, one on link 1 and one on link 2, which must always have the same velocity if they are to be connected. By the same reasoning, instant center $(2, 3)$ is located at the connection between links 2 and 3. Link 3 must travel in a straight line. Therefore, if link 3 is to rotate about a center located on link 4, that center must be located at infinity in a direction perpendicular to the direction of travel of link 3.

Considering links 1, 2, and 3 there will be three instant centers $(1, 2)$, $(2, 3)$, and $(1, 3)$. Since the locations of instant centers $(1, 2)$ and $(2, 3)$ are known, a straight line through them will pass through instant center $(1, 3)$. Considering links 1, 3, and 4 there will be three instant centers $(1, 4)$, $(1, 3)$, and $(3, 4)$. A straight line through instant center $(1, 4)$ and perpendicular to the direction of travel of link 3 will pass through instant centers $(1, 3)$ and $(3, 4)$. Since instant center $(1, 3)$ must be located on both straight lines, it is located at their intersection.

A straight line through instant centers $(1, 4)$ and $(1, 2)$ will pass through instant center $(4, 2)$. Another straight line through instant center $(2, 3)$ and perpendicular to the direction of travel of link 3 will pass through instant centers $(3, 4)$ and $(2, 3)$, thus locating instant center $(2, 4)$.

For more complex mechanisms as shown in Fig. 3.8, location of the instant centers for velocity is enhanced by lightly drawing a circle and locating points with numbers corresponding to the number of each link on the circle. When an instant center is located, a line is drawn between the two numbers indicating that the instant center has been located. When each numbered point on the circle is connected to each other point by a straight line, all instant centers have been located. More important,

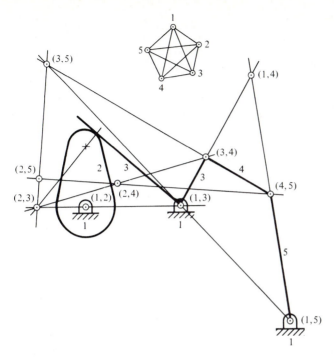

FIGURE 3.8
Location of instant centers.

selecting three links whose three instant centers lie on a straight line results in selecting a triangle of numbered points. Therefore, a triangle in the circle of numbers represents a straight line on the mechanism drawing.

In Fig. 3.8, instant centers $(1, 2)$, $(1, 3)$, $(3, 4)$, $(4, 5)$, and $(1, 5)$ are located by inspection. Link 2 is a cam which slides along link 3 which is straight, and the instant center $(2, 3)$ must therefore be located along the perpendicular to link 3 through the point of contact of links 2 and 3. Instant center $(2, 3)$ must also be located on the straight line through instant centers $(1, 2)$ and $(1, 3)$.

Select triangle 1, 4, 3 and triangle 1, 4, 5 from the circle of numbers because side 1, 4 is common to the two triangles and represents instant center $(1, 4)$. Instant center $(1, 4)$ is then located at the intersection of a straight line through instant centers $(1, 3)$ and $(3, 4)$ and a straight line through instant centers $(3, 5)$ and $(4, 5)$. By selecting two triangles with a common side as the unknown instant center location, all remaining instant centers can be located.

Quite often the instant center is located far off the page. When it is necessary to draw a line to such a center, the procedure described in Fig.

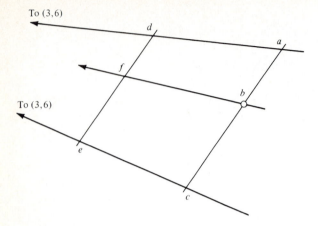

FIGURE 3.9
Construction for remote centers.

3.9 is convenient. If a line is to be drawn from point b to instant center $(3, 6)$, an arbitrary line is drawn through point b and intersects two rays to $(3, 6)$ at points a and c. Another line is drawn parallel to the line abc and intersects the two rays at d and e. The line df is divided such that $df : de = ab : ac$. Then a line through b and f will also pass through center $(3, 6)$.

3.4 ANALYSIS BY USE OF INSTANT CENTERS FOR VELOCITY

The Whitworth quick-return mechanism of Fig. 3.10 is designed with the input crank (link 1) rotating counterclockwise at a known rate of radians per second. The drawing is made to scale so that any dimensions may be simply scaled from the drawing. The linear velocity of link 6 and the angular velocity of link 5 are both desired.

All 15 instant centers for velocity are located either by inspection or by application of the Aronhold-Kennedy rule.

To find the linear velocity of link 6, it is only necessary to find the velocity of one point on link 6 since all points on the link must have the same velocity if the link is to move in a straight line. Since link 1 is the driver with known angular velocity, the linear velocity of any point on link 1 is simply the distance from that point to instant center $(1, 4)$ multiplied by the angular velocity of link 1. Instant center $(1, 6)$ represents a point on link 1 coincident with a point on link 6, and those two points have the same velocity. The velocity of instant center $(1, 6)$ considered as a point on link 1 is equal to the distance from center $(1, 6)$ to center $(1, 4)$ multiplied by the angular velocity of link 1. The linear

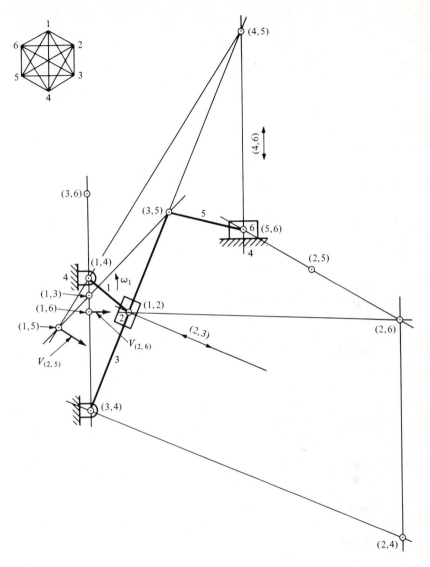

FIGURE 3.10
Velocity analysis using instant centers.

velocity of link 6 is then equal to the linear velocity of instant center $(1, 6)$. Direction of the velocity of link 6 is determined by the sense of rotation of link 1.

To find the angular velocity of link 5, it is necessary to know the linear velocity of a point on link 5. It would be convenient to calculate the linear velocity of instant center $(1, 5)$ on link 1 and remember that that point on link 5 has the same velocity. It is more convenient to realize that instant center (5, 6) must have the same velocity as instant center (1, 6) since all points on link 6 must have the same velocity. Using the velocity of instant center (5, 6) and dividing by the distance from instant center (5, 6) to instant center (4, 5), the angular velocity of link 5 may be determined. The sense of the angular velocity of link 5 is found by the direction of the velocity of instant center (5, 6) and realizing that the absolute motion of link 5 consists of rotation about center (4, 5) at this instant.

3.5   POLODES

A *polode* (sometimes known as a centrode) represents the locus of all possible positions of the instant center for velocity of one link with respect to another link. In Fig. 3.11, the polode representing possible locations of the instant center for velocity (2, 4) is determined by rotating link 1 into various possible positions and finding the resulting position of instant center (2, 4). It is interesting to note that instant center (2, 4) is a point on link 2 which has the same velocity as a coincident point on link 4. All points on link 4 have zero velocity which means that instant center (2, 4) represents a point on link 2 which has zero velocity. However, the instant center indicates the state of the mechanism at an instant. This means that, at this particular instant, the instant center considered as a point on link 2 has zero velocity. At a later instant, the point on link 2 which has zero velocity may be located at a different site on link 4. Thus the locus of the instant center (2, 4) must exist. The polode [locus of instant center (2, 4)] is called the *fixed polode* and given the symbol π_f.

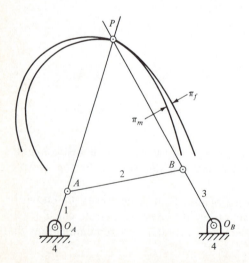

FIGURE 3.11
Fixed and moving polodes.

If now the mechanism is inverted (which does not change any of the relative motions of the mechanism) and the polode of instant center $(2, 4)$ is again located, that polode is called the *moving polode* and given the symbol π_m. The fixed and moving polodes between links 2 and 4 are shown in Fig. 3.11. The point of contact of the fixed and moving polodes is generally called the *pole point*.

If a sheet of plywood is cut out in the shape of the fixed polode (π_f) and another piece is cut out in the shape of the moving polode (π_m) and placed in contact at the pole point P, an interesting concept results. Now, rolling (without slipping) the moving polode along the fixed polode will cause points A and B (scribed on the moving polode) of the original coupler link to move in circular arcs as though they were rotating about their fixed centers O_A and O_B. This means that motion of the coupler plane which includes AB can be simulated by rolling the moving polode on the fixed polode. This is an important concept in synthesis of mechanisms.

Radius of curvature of the polodes at their point of contact and variation of the radius of curvature has been considered in detail and is reported in the literature.

3.6   THE VELOCITY POLYGON

The velocity polygon technique provides a quick graphical means of determination of the velocity of any point in a mechanism. In addition, it provides the relative velocities needed for acceleration analysis. The technique relies on the principle that the velocity vector is directed perpendicular to a link of fixed length and in the direction indicated by the angular velocity of the link. The technique also relies on the fact that the velocity of one point on a link related to any other point on the same link is directed perpendicular to the link and is equal to the distance between the two points multiplied by the angular velocity of the link. In Fig. 3.12, the velocity of point B is equal to the velocity of point A plus (vectorially) the velocity of point B with respect to point A. The vector polygon used to complete the addition is referred to as the velocity polygon and is drawn to scale [one unit of length on the drawing represents x units per second (units/s) velocity] to allow measurement of the results. The polygon is shown centered at B but may be placed anywhere on the page. If the length of the link, the velocity of point A, and the angular velocity of link 3 are all known, the velocity of point B in Fig. 3.12 is found from the velocity polygon.

The equation to be solved is

$$V_B = V_A + V_{B/A}$$

Note that all absolute velocity vectors for a mechanism originate from the center of the velocity polygon.

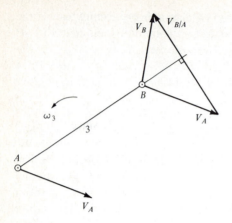

FIGURE 3.12
The velocity polygon.

In Fig. 3.12, the velocity of A and angular velocity of link 2 are known. The velocity of B with respect to A is equal to the distance AB multiplied by ω_3 and is directed perpendicular to AB in the sense indicated by ω_3.

In the mechanism shown in Fig. 3.13, link 1 is rotating counterclockwise at 1500 revolutions per minute (r/min) $[1500/2\pi = 238.7$ radians per minute (rad/min) or 3.98 radians per second (rad/s)]. Dimen-

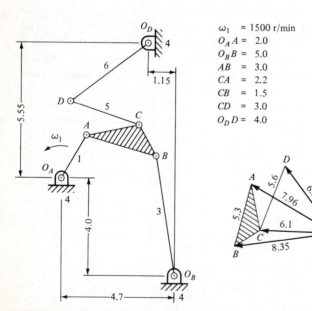

$$\omega_1 = 1500 \text{ r/min}$$
$$O_A A = 2.0$$
$$O_B B = 5.0$$
$$AB = 3.0$$
$$CA = 2.2$$
$$CB = 1.5$$
$$CD = 3.0$$
$$O_D D = 4.0$$

FIGURE 3.13
Velocity analysis with the velocity polygon.

sions of the components are in U.S. conventional or metric units and are referred to as units. The velocity polygon is constructed as follows:

$$V_A = O_A A(\omega_1) = 2(3.98) = 7.96 \text{ units/s} \qquad \text{up and to the left perpendicular to link 1}$$

$$V_{B/A} = AB(\omega_2) \qquad \text{directed perpendicular to link 2}$$

$$V_B = V_A + V_{B/A} \qquad \text{directed perpendicular to link 3 since } B \text{ is also a point on link 3 and its velocity must be directed perpendicular to the link}$$

From the polygon

$$V_B = 8.35 \text{ units/s}$$

and $\qquad V_{B/A} = 5.30 \text{ units/s}$

or $\qquad \omega_2 = \dfrac{V_{B/A}}{AB} = \dfrac{5.30}{3} = 1.77 \text{ rad/s} \qquad$ clockwise (CW) since the velocity of B/A is down

The velocity of point C is found by simultaneous solution of two equations:

$$V_C = V_A + V_{C/A} \qquad V_C = V_B + V_{C/B}$$

$$V_C = 6.10 \text{ units/s} \qquad \text{scaled from the polygon}$$

This is possible since points A, B, and C are all points on the same link. The velocity $V_{C/A}$ is directed perpendicular to AC, and the velocity $V_{C/B}$ is directed perpendicular to BC. With this construction, the triangle ABC in the velocity polygon becomes similar to the triangle ABE in the configuration drawing of the mechanism and is known as the *image* of the coupler link ABC.

The velocity of point D is known to be directed perpendicular to link 6. The velocity of point D is also equal to the velocity of point C plus the velocity of point D with respect to point C since C and D are two points on the same link. In the velocity polygon, the velocity $V_{D/C}$ is made perpendicular to CD and the velocity of point D is made perpendicular to link 6. Then

$$V_D = 6.80 \text{ units/s} \qquad \text{scaled from the polygon}$$

$$V_{C/D} = 5.60 \text{ units/s} \qquad \text{scaled from the polygon}$$

and $\qquad \omega_6 = \dfrac{V_D}{O_D D} = \dfrac{6.8}{4} = 1.70 \text{ rad/s CW}$

$$\omega_5 = \dfrac{V_{D/C}}{CD} = \dfrac{5.6}{3} = 1.87 \text{ rad/s CW}$$

These results apply only when the mechanism is in the phase shown. A short time later, the phase of the mechanism will be different and different values for the velocities will be found.

3.7   USE OF AN AUXILIARY POINT

For some complex mechanisms such as that shown in Fig. 3.14 (the Stephenson linkage), it is not possible to construct a velocity polygon directly. With the angular velocity of link 2 known and thus the velocity of points A and B known, the velocity of any point on link 4 is desired. Since the direction of the velocity of points C and D is unknown, direct construction of the velocity polygon is not possible.

Point G (Fig. 3.14) is an auxiliary point located by extending links 3 and 5 to their intersection. (Other auxiliary points may be used if the relative velocities are related as follows:). Point G is considered to be a point on link 4. With the velocity of points A and B known, the procedure is as follows:

$$V_C = V_A + V_{C/A}$$

In this equation, the velocity of A is known and the direction of the velocity of C with respect to A is known but nothing is known about the velocity of C. Now the velocity of G can be found:

$$V_G = V_C + V_{G/C} = V_A + (V_{C/A} + V_{G/C})$$

also $$V_G = V_D + V_{G/D} = V_B + (V_{D/B} + V_{G/D})$$

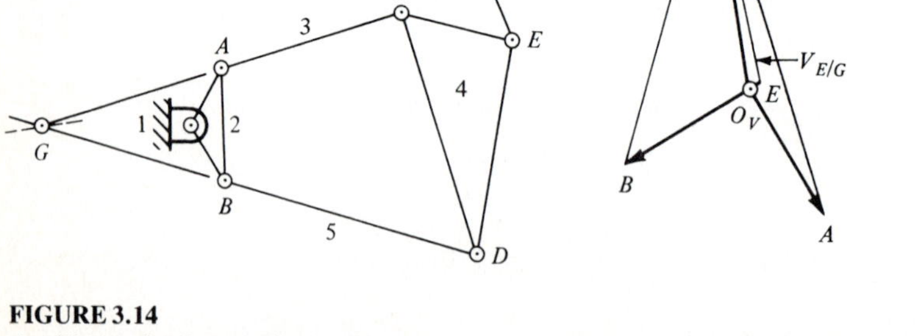

FIGURE 3.14
Auxiliary point for velocity analysis.

Although the magnitudes of $(V_{C/A} + V_{G/C})$ and each of the relative velocities are not known, the direction of their sum is perpendicular to the line AC or CG. Thus, the intersection of $(V_{C/B} + V_{G/C})$ drawn from the head of V_A and $(V_{D/B} + V_{G/D})$ drawn from the head of V_B will represent a simultaneous solution for the two equations describing V_G.

With the velocity of point G known, the velocity of point E may be found since its direction is known to be perpendicular to link 6 and

$$V_E = V_G + V_{E/G}$$

3.8 ANALYTIC VELOCITY ANALYSIS

If the link of a mechanism is expressed as a vector in the form $qe^{i\theta}$, differentiation of the vector will yield

$$d(qe^{i\theta}) = qe^{i\theta}(i\,d\theta) + e^{i\theta}[d(q)]$$

Since the length of the link is constant, $d(q) = 0$. If the differentiation is with respect to time,

$$d(qe^{i\theta}) = iq\omega e^{i\theta}$$

where ω represents angular velocity and i indicates 90° rotation in the direction of ω.

In Fig. 3.15, point B may be located by one of two loop equations which must be equal.

$$O_A B = r_1 e^{i\theta_1} + r_2 e^{i\theta_2}$$

$$O_A B = r_4 + r_3 e^{i\theta_3}$$

or

$$r_1 e^{i\theta_1} + r_2 e^{i\theta_2} = r_4 + r_3 e^{i\theta_3}$$

Differentiation with respect to time results in

$$ir_1 \omega_{r1} e^{i\theta_1} + ir_2 \omega_{r2} e^{i\theta_2} = 0 + ir_3 \omega_{r3} e^{i\theta_3}$$

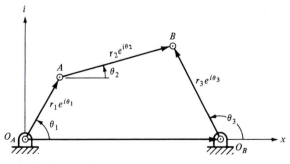

FIGURE 3.15
General open four-bar mechanism.

Dividing through by i and expressing each term of the equation in terms of its equivalent sin and cos form gives

$$r_1\omega_{r1}(\cos\theta_1 + i\sin\theta_1) + r_2\omega_{r2}(\cos\theta_2 + i\sin\theta_2)$$
$$= r_3\omega_{r3}(\cos\theta_3 + i\sin\theta_3)$$

This equation may be separated into its real and imaginary components to produce two separate equations.

$$r_2\omega_{r2}\cos\theta_2 - r_3\omega_{r3}\cos\theta_3 = -r_1\omega_{r1}\cos\theta_1 \qquad (a)$$

$$r_2\omega r_2\sin\theta_2 - r_3\omega_{r3}\sin\theta_3 = -r_1\omega_{1r1}\sin\theta_1 \qquad (b)$$

In these two equations, the only unknowns are ω_{r2} and ω_{r3}. The two equations may be solved simultaneously by subtracting one from the other to eliminate one of the unknowns, or they may be solved using Cramer's rule.

Using Cramer's rule:

$$\omega_{r2} = \frac{\begin{vmatrix} -r_1\omega_{r1}\cos\theta_1 & -r_3\cos\theta_3 \\ -r_1\omega_{r1}\sin\theta_1 & -r_3\sin\theta_3 \end{vmatrix}}{\begin{vmatrix} r_2\cos\theta_2 & -r_3\cos\theta_3 \\ r_2\sin\theta_2 & -r_3\sin\theta_3 \end{vmatrix}}$$

$$= \frac{-r_1r_3\omega_{r1}\cos\theta_3\sin\theta_1 + r_1r_3\omega_{r1}\sin\theta_3\cos\theta_1}{-r_2r_3\sin\theta_3\cos\theta_2 + r_2r_3\sin\theta_2\cos\theta_3}$$

$$= \omega_{r1}\frac{r_1}{r_2}\frac{\sin(\theta_1 - \theta_3)}{\sin(\theta_3 - \theta_2)} \qquad (3.2)$$

Using algebraic manipulation rather than Cramer's rule to solve for ω_{r3} multiply Eq. (a) by $\sin\theta_2$, multiply equation (b) by $\cos\theta_2$, and subtract the resulting equations.

$$r_1\omega_{r1}\cos\theta_1\sin\theta_2 = r_3\omega_{r3}\cos\theta_3\sin\theta_2 - r_2\omega_{r2}\cos\theta_2\sin\theta_2$$
$$\underline{r_1\omega_{r1}\sin\theta_1\cos\theta_2 = r_3\omega_{r3}\sin\theta_3\cos\theta_2 - r_2\omega_{r2}\cos\theta_2\sin\theta_2}$$
$$r_1\omega_{r1}(\cos\theta_1\sin\theta_2 - \sin\theta_1\cos\theta_2) = r_3\omega r_3(\cos\theta_1\sin\theta_2 - \sin\theta_1\cos\theta_2)$$

$$\omega_{r3} = \omega_{r1}\frac{r_1}{r_3}\frac{\sin(\theta_1 - \theta_2)}{\sin(\theta_3 - \theta_2)} \qquad (3.3)$$

3.9   COMPUTER PROGRAMS FOR VELOCITY ANALYSIS

Equations (3.2) and (3.3) may be solved using a computer program. Output from the program VELOCITY in App. A will be

* * * * * * * * * * * * * * * VELOCITY * * * * * * * * * * * * * * * * * *

THIS PROGRAM PROVIDES ANGULAR VELOCITY OF EACH LINK
(COUNTERCLOCKWISE POSITIVE) OF AN OPEN OR CROSSED FOUR-BAR
MECHANISM. CRANK ANGLE IN DEGREES. VELOCITIES IN RADIANS PER
SECOND.

LENGTH OF FIXED LINK   = 10
LENGTH OF CRANK   = 4
LENGTH OF COUPLER   = 12
LENGTH OF ROCKER   = 8
CRANK ANGULAR VELOCITY = 4

MECHANISM IS OPEN

CRANK ANGLE	COUPLER ANGULAR VELOCITY	ROCKER ANGULAR VELOCITY
0	−2.666666	−2.666666
30	−1.048963	.486932
60	− .158227	1.869388
90	.257070	2.155924
120	.557829	2.057255
150	.854830	1.694080
180	1.142849	1.142877
210	1.331706	.492506
240	1.288328	− .211063
270	.846408	−1.052430
300	− .262750	−2.250372
330	−2.143028	−3.678981
360	−2.666666	−2.666666

If the velocity of a coupler point is desired, the above program may
be modified slightly (CPVELOCITY, App. A). With reference to Fig.
3.16, the distance R1 is measured along the coupler link from its

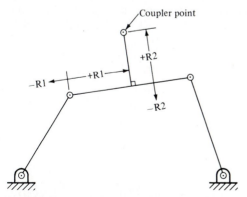

FIGURE 3.16
Coupler point location.

connection with the input crank and the distance R2 is measured perpendicular from the coupler link to the coupler point. Positive and negative values of R1 and R2 are indicated in Fig. 3.16.

PROBLEMS

3.1. An automobile with 28-inch (in) outside diameter tires is moving at 55 mi/h. Determine
(*a*) The angular velocity of the wheel in revolutions per minute.
(*b*) The linear velocity of a point on the top of the tire.

3.2. The hour hand of a town clock is 7 ft long and the minute hand is 5 ft long. What is the relative velocity of the tips of the hands in inches per minute at 3 o'clock?

3.3. In the figure, gear 2 rotates at 1750 r/min. Find the velocity of point *P* and of point *A* in inches per minute.

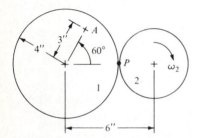

FIGURE P3.3
Meshing gears.

3.4. A piece of material with a diameter of 8 in has a recommended cutting speed of 120 ft/min. What should be the angular velocity of the material in revolutions per minute when being turned on a lathe?

3.5. In a centrifugal fan the relative velocity of a gas particle with respect to the blade is 97 meters per second (m/s). If the fan is rotating at 1720 r/min, determine the blade tip velocity and the particle absolute velocity when leaving the blade.

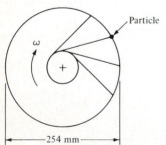

FIGURE P3.5
Centrifugal fan.

3.6. Using the method of components determine the linear velocity of points B, C, and D.

$$\omega_1 = 4.00 \text{ rad/s} \qquad O_A A = 1.50 \qquad O_B B = 1.25$$

$$AB = 2.25 \qquad\qquad AC = 1.25 \qquad BC = 2.50 \qquad CD = 3.00$$

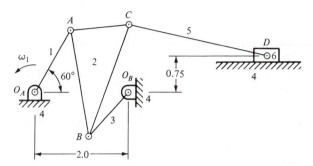

FIGURE P3.6
Method of comonents.

3.7. Using the method of components determine the linear piston velocity and the velocity of point B of the Andreau differential stroke engine.

$$\omega_1 = 10.0 \text{ rad/s} \qquad O_A A = 0.75 \qquad O_B B = 1.50$$

$$AC = 3.00 \qquad\qquad BC = 3.00 \qquad CD = 4.25$$

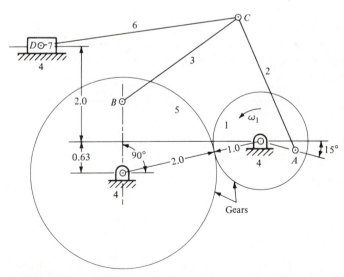

FIGURE P3.7
Method of components.

3.8. (*a*) Locate all the instant centers for velocity.
  (*b*) If $\omega_1 = 3$ rad/s CCW, determine ω_2, ω_3, ω_5, ω_6.

$$O_A A = 1.50 \qquad O_B B = 1.00 \qquad O_D D = 2.50$$
$$AB = 3.00 \qquad AC = 2.00 \qquad BC = 1.75 \qquad CD = 1.00$$

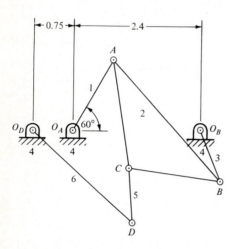

FIGURE P3.8
Instant centers for velocity location.

3.9. (*a*) Locate all the instant centers for velocity.
  (*b*) Determine V_C, ω_2, ω_3, ω_5 with $\omega_1 = 4$ rad/s CW.

$$O_B A = 2.00 \qquad O_B B = 1.00 \qquad BC = 2.50$$

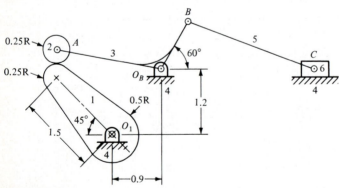

FIGURE P3.9
Instant centers for velocity location.

3.10. Carefully plot the fixed and moving polodes associated with link 2. Use θ_1 from 0 to 180° in 15° increments.

$$O_A O_B = 3.5 \qquad O_A A = 1.0 \qquad O_B B = 2.0 \qquad AB = 2.0$$

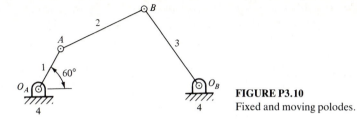

FIGURE P3.10
Fixed and moving polodes.

3.11. Carefully plot the fixed and moving polodes associated with link 2.

$$OA = OB = 2.0$$

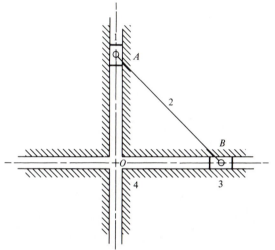

FIGURE P3.11
Fixed and moving polodes.

3.12. Construct a velocity polygon and determine V_B, V_C, V_D, ω_2, ω_3, and ω_5. Check using a computer program similar to the one in Sec. 3.9.

$O_A A = 1.00$	$O_B B = 1.50$	$AB = 3.00$	$\omega_1 = 1500 \, \text{r/min}$
$AC = 1.75$	$BC = 1.75$	$CD = 3.25$	

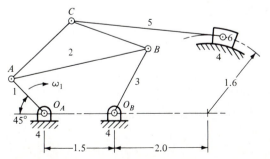

FIGURE P3.12
Velocity polygon.

3.13. Construct a velocity polygon for the Atkinson engine. If the piston is moving to the left at 4 ft/s, determine ω_1 and ω_3.

$$O_A A = 2.00 \text{ in} \qquad O_C C = 2.00 \text{ in} \qquad AC = 4.25 \text{ in}$$
$$CB = 4.90 \text{ in} \qquad AB = 1.00 \text{ in} \qquad BD = 4.00 \text{ in}$$

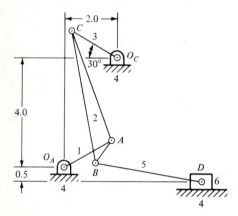

FIGURE P3.13
Velocity polygon.

3.14. With $\omega_1 = 35 \text{ rad/s}$, determine the linear velocity of points B and C and then the angular velocities of links 3 and 5.

$$O_A A = 1.0 \qquad O_B B = 2.5 \qquad O_B A = 1.5 \qquad BC = 2.0$$

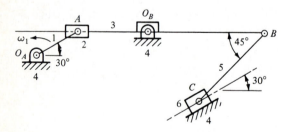

FIGURE P3.14
Velocity polygon.

3.15. Link 1 rotates at 50 r/min CW. Determine the linear velocity of points B and C and the angular velocities of links 2, 3, and 5.

$$O_A O_B = 6.5 \qquad O_A A = 2.0 \qquad O_B B = 2.0$$
$$AB = 4.0 \qquad BC = 4.0$$

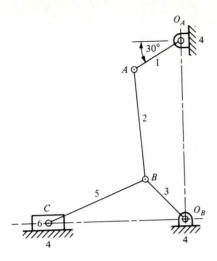

FIGURE P3.15
Angular velocity by polygon.

3.16. Link 1 is rotating in a CCW sense at 125 r/min. Determine the linear velocity of point F and the angular velocities of links 5, 6, and 7.

$$O_A O_D = 10 \qquad O_A A = 2 \qquad O_D D = 6 \qquad AB = 5 \qquad AC = 3$$
$$BC = 4 \qquad CD = 3 \qquad DE = 4 \qquad EF = 4$$

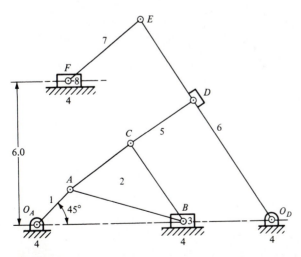

FIGURE P3.16
Linear and angular velocity by polygon.

3.17. Find the angular velocity of the floating link *BCE*.

$$BC = 3.25 \qquad CE = 2.91 \qquad BE = 1.60$$
$$AB = CD = DE = 1.0$$
$$V_A = V_D = V_F = 15 \text{ units/s}$$

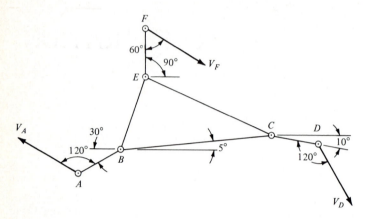

FIGURE P3.17
Floating link.

CHAPTER
4

ACCELERATION

4.1 CONCEPTS

Acceleration is the time rate of change of velocity. In considering acceleration it is necessary to recognize that the effective length of a link is constant or may be changing as in the case of a link along which a slider is moving. Notice that it is the effective length and not the actual length of the link that is considered. In the case of a link connected with a revolute joint to other links, the actual length of the link is the effective length. In the case of a link connected to another link with a sliding (prismatic) joint, the effective length is not the actual length and more careful consideration is necessary. Mechanisms in which the effective length is the actual length of the link will be considered first. Later, the case of changing effective length of the link will be considered leading to the Coriolis acceleration component.

In Fig. 4.1 a link of fixed length is shown rotating about a fixed center. The link has an angular velocity of ω_1 at time 1 and an angular velocity of ω_2 at time 2. The length of the link is R. At time 1, the velocity of point E is $V_{E1} = R\omega_1$ directed perpendicular to the link in phase 1. At time 2, the velocity of point E is $V_{E2} = R\omega_2$ directed perpendicular to the link in phase 2.

The two velocities may be assembled into a velocity polygon to define the change in velocity or $\Delta V_E = V_{E2} - V_{E1}$. Point C is located so that $O_V C = O_V B$. The change in velocity ΔV_E may then be represented as the vector sum of **BC** and **CD**. Then the change in velocity $\Delta V_E =$

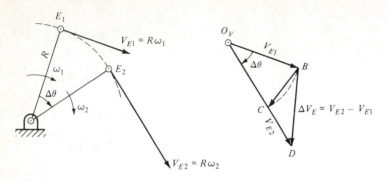

FIGURE 4.1
Acceleration of point E.

BC + CD. Acceleration is the time rate of change of velocity. Then the acceleration $\mathbf{A}_E$ becomes

$$\mathbf{A}_E = \lim_{\Delta t \to 0} \frac{\Delta V}{\Delta t} = \lim_{\Delta t \to 0} \left(\frac{BC}{\Delta t} + \frac{CD}{\Delta t} \right)$$

Considering first the limit as $\Delta t \to 0$ of $BC/\Delta t$ recall that the distance $O_V B$ is equal to the velocity $\mathbf{V}_{E1}$. The distance BC then becomes $\mathbf{V}_{E1}(\Delta \theta)$ and

$$\lim_{\Delta t \to 0} \frac{BC}{\Delta t} = \lim_{\Delta t \to 0} \frac{V \, \Delta \theta}{\Delta t} = \frac{V \, d\theta}{dt} = V\omega$$

Since $V = R\omega$

$$\lim_{\Delta t \to 0} \frac{BC}{\Delta t} = V\omega = R\omega^2 = \frac{V^2}{R} \tag{4.1}$$

Note that as the time becomes very small and $\Delta \theta \to 0$, the vector **BC** becomes perpendicular to the velocity vector $\mathbf{V}_{E1}$ or is parallel to the link and directed toward the center of rotation of the link. Since its direction is perpendicular to the velocity vector or perpendicular to the path of point E, this acceleration component is known as the *normal acceleration.*

The velocity of a point is always the angular velocity of the radius of curvature of the path of the point multiplied by the distance to the center of curvature. Then the distance $CD = R\omega_2 - R\omega_1 = R \, \Delta\omega$. Then

$$\lim_{\Delta t \to 0} \frac{CD}{\Delta t} = \lim_{\Delta t \to 0} R \frac{\Delta\omega}{\Delta t} = R \frac{d\omega}{dt} = R\alpha \tag{4.2}$$

where α is the angular acceleration of the link.

In the limit as $\Delta\theta \to 0$ the vector $R\alpha$ becomes perpendicular to the vector $R\omega^2$ and is directed as indicated by the changing angular velocity of the link. Since the vector $R\alpha$ is directed tangent to the path of point E, it is known as the *tangential acceleration.*

Thus it is seen that the acceleration of a point on a link of fixed length which is rotating at a changing angular velocity is equal to the vector sum of the normal and the tangential acceleration components or

$$\mathbf{A} = \mathbf{A}^n = \mathbf{A}^t$$

4.2 GRAPHICAL ACCELERATION ANALYSIS

A graphical acceleration analysis may be conducted using components of the acceleration or by using the velocity polygon. Use of the velocity polygon is more convenient and is recommended.

The six-bar mechanism of Fig. 4.2 is to be analyzed to determine the acceleration of points of interest on the mechanism and to determine the angular acceleration of each link of the mechanism. In finding the angular acceleration, it is essential that the direction of the acceleration be determined as well as its magnitude. In later analysis of inertial forces, the direction of the angular acceleration plays a very important role. In Fig. 4.2, the crank (link 1) is rotating counterclockwise and increasing in angular velocity. The length of all links is known. The analysis proceeds as follows:

$$\mathbf{V}_A = O_A A(\omega_1) = (1.5)(3.0) = 4.50 \text{ units/s}$$

With the direction of the velocity of points B and D known to be perpendicular to links 3 and 6, the velocity polygon may be drawn as shown. Since the polygon is drawn to scale, the velocity of any point or any relative velocity may be scaled from the polygon.

The normal acceleration of point $A = O_A A(\omega_1^2)$. Then

$$\mathbf{A}_A^n = (1.5)(3^2) = 13.5 \text{ units/s}^2$$

The normal acceleration of point A is directed from point A toward its center of rotation O_A.

The tangential acceleration of of point $A = O_A A \alpha_1$. Then

$$\mathbf{A}_A^t = (1.5)(10.0) = 15.0 \text{ units/s}^2$$

The tangential acceleration of point A is directed perpendicular to the normal acceleration of point A and in the direction indicated by the direction of α_1.

The normal acceleration of point B is directed from B toward its center of rotation O_B and is equal to the velocity of B squared divided by the radius of curvature of the path of point B or

$$\mathbf{A}_B^n = \frac{V_B^2}{O_B B} = \frac{(3.0)^2}{1.75} = 5.14 \text{ units/s}^2$$

The tangential acceleration of point B is directed perpendicular to

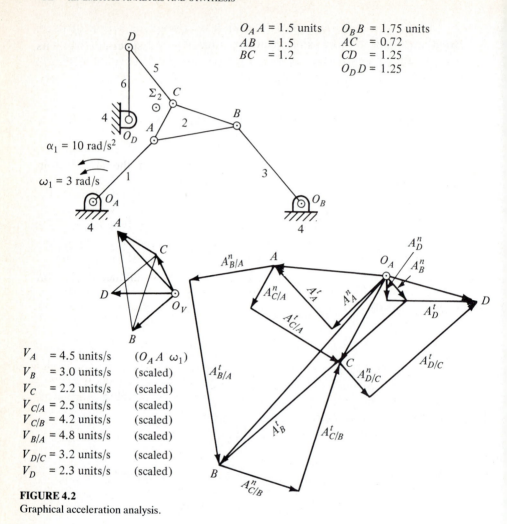

$O_A A$ = 1.5 units	$O_B B$ = 1.75 units
AB = 1.5	AC = 0.72
BC = 1.2	CD = 1.25
	$O_D D$ = 1.25

α_1 = 10 rad/s²

ω_1 = 3 rad/s

V_A = 4.5 units/s	$(O_A A\ \omega_1)$	
V_B = 3.0 units/s	(scaled)	
V_C = 2.2 units/s	(scaled)	
$V_{C/A}$ = 2.5 units/s	(scaled)	
$V_{C/B}$ = 4.2 units/s	(scaled)	
$V_{B/A}$ = 4.8 units/s	(scaled)	
$V_{D/C}$ = 3.2 units/s	(scaled)	
V_D = 2.3 units/s	(scaled)	

FIGURE 4.2
Graphical acceleration analysis.

the normal acceleration of point B, but its direction is unknown since the sense of the angular acceleration of link 4 is unknown.

The normal acceleration of point B with respect to point A is equal to the velocity of point B with respect to point A squared divided by the distance AB. The velocity of B with respect to A is available from the velocity polygon.

$$\mathbf{A}^n_{B/A} = \frac{V^2_{B/A}}{AB} = \frac{(4.8)^2}{1.5} = 15.36 \text{ units/s}^2$$

The normal acceleration of B with respect to A is directed from B toward A.

The tangential acceleration of B with respect to A is directed perpendicular to the normal acceleration of B with respect to A, but its direction is unknown since the sense of the angular acceleration of link 2 is unknown. However, now two expressions for the acceleration of point B are known and can be solved simultaneously.

$$\mathbf{A}_B = \mathbf{A}_A + \mathbf{A}^n_{B/A} + \mathbf{A}^t_{B/A}$$

$$\mathbf{A}_B = \mathbf{A}^n_B + \mathbf{A}^t_B$$

These two equations are solved graphically in the acceleration polygon yielding

$\mathbf{A}_B = 44.2$ units/s^2

$\mathbf{A}^t_B = 44.0$ units/s^2 directed down and to the left

$\mathbf{A}^t_{B/A} = 32.5$ units/s^2 directed down and to the right

The acceleration of point C may be found by simultaneous solution of two equations:

$$\mathbf{A}_C = \mathbf{A}_A + \mathbf{A}^n_{C/A} + \mathbf{A}^t_{C/A}$$

$$\mathbf{A}_C = \mathbf{A}_B + \mathbf{A}^n_{C/B} + \mathbf{A}^t_{C/B}$$

The normal acceleration of C with respect to A is equal to the velocity of C with respect to A squared and divided by the distance CA. The velocity of C with respect to A is available from the velocity polygon.

$$\mathbf{A}^n_{C/A} = \frac{2.5^2}{0.72} = 8.68 \text{ units/s}^2$$

The tangential acceleration of C with respect to A is directed perpendicular to the normal acceleration of C with respect to A.

The normal acceleration of C with respect to B is equal to the velocity of C with respect to B squared and divided by the distance CB.

$$\mathbf{A}^n_{C/B} = \frac{4.2^2}{1.2} = 14.7 \text{ units/s}^2$$

The tangential acceleration of C with respect to B is directed perpendicular to the normal acceleration of C with respect to B. Intersection of the two tangential acceleration vector directions locates the acceleration of point C as well as the magnitude and direction of the tangential accelerations of C with respect to A and B.

$\mathbf{A}_C = 17.0$ units/s^2

$\mathbf{A}^t_{C/A} = 18.2$ units/s^2 directed down and to the right

$\mathbf{A}^t_{C/B} = 23.6$ units/s^2 directed up and to the right

The acceleration of point D is found by simultaneous solution of two equations:

$$\mathbf{A}_D = \mathbf{A}_C + \mathbf{A}^n_{D/C} + \mathbf{A}^t_{D/C} \qquad \text{with } D \text{ considered as a point on link 5}$$

$$\mathbf{A}_D = \mathbf{A}^n_D + \mathbf{A}^t_D \qquad \text{with } D \text{ considered as a point on link 6}$$

The normal acceleration of D/C is equal to the velocity of D with respect to C (from the velocity polygon) squared divided by the distance CD and is directed from D toward C. The normal acceleration of point D on link 6 is equal to the velocity of point D (from the velocity polygon) squared divided by the distance $O_D D$ and is directed from D toward O_D.

The tangential acceleration of D with respect to C and the tangential acceleration of D as a point on link 6 are found from the acceleration polygon.

$\mathbf{A}^t_{D/C} = 26.0$ units/s^2 directed up and to the right

$\mathbf{A}^t_D \;\; = 17.0$ units/s^2 directed to the right

Now the angular acceleration of each link may be found since the tangential acceleration of a point is equal to the distance from its center of rotation multiplied by the angular acceleration of the link.

$$\alpha_2 = \frac{\mathbf{A}^t_{B/A}}{AB} = \frac{32.5}{1.5} = 21.66 \text{ rad/s}^2 \text{ CW}$$

$$\alpha_3 = \frac{\mathbf{A}^t_B}{O_B B} = \frac{44.0}{1.75} = 25.14 \text{ rad/s}^2 \text{ CCW}$$

$$\alpha_5 = \frac{\mathbf{A}^t_{D/C}}{CD} = \frac{26.0}{1.25} = 20.80 \text{ rad/s}^2 \text{ CW}$$

$$\alpha_6 = \frac{\mathbf{A}^t_D}{O_D D} = \frac{17.0}{1.25} = 13.60 \text{ rad/s}^2 \text{ CW}$$

The sense of the angular acceleration (CW or CCW) is determined by the direction of the tangential acceleration as indicated on the acceleration polygon. It is important that the sense of the angular acceleration be correctly determined for later use in calculation of inertial forces and dynamic balancing.

In the acceleration polygon, the triangle formed by the terminus of the acceleration of points A, B, and C forms a triangle which is similar to the triangle ABC on the configuration drawing. The triangle in the acceleration polygon is known as the image of the triangle on the configuration drawing.

Using the image concept, the triangle ABO_A of the acceleration polygon may be transferred to the configuration diagram to locate that point on link 2 which has zero acceleration. The point of zero acceleration on the configuration drawing is the location of the instant center for acceleration for link 2 (Σ_2). Note that the instant center for acceleration and the instant center for velocity are two different points on the configuration drawing.

4.3 ACCELERATIONS IN ROLLING CONTACT

In the case of one body rolling on another body without slipping as shown in Fig. 4.3, the acceleration of the point of contact is of interest. In Fig. 4.3 the point of contact P is recognized as the instant center for velocity of link 2 with respect to link 1. The instant center for velocity has zero velocity in this case, but its acceleration is not zero. The acceleration of point P on link 2 can be found from the acceleration of point C on link 2. Since there is no slipping, the angular acceleration α_2 is equal to zero resulting in $\mathbf{A}^t_{P/C} = \mathbf{0}$ and $\mathbf{A}^t_C = \mathbf{0}$ in both cases. The path of point C is a circle of radius $(r_1 + r_2)$ or $(r_1 - r_2)$.

$$\mathbf{A}^n_C = \frac{\mathbf{V}^2_C}{r_1 + r_2} \qquad \mathbf{A}^n_C = \frac{\mathbf{V}^2_C}{r_1 - r_2}$$

With P as the instant center for velocity $\mathbf{V}_C = r_2 \omega_2$

$$\mathbf{A}^n_C = \frac{r_2^2 \omega_2^2}{r_1 + r_2} \qquad \mathbf{A}^n_C = \frac{r_2^2 \omega_2^2}{r_1 - r_2}$$

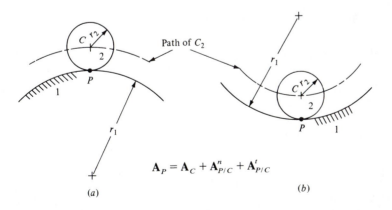

$$\mathbf{A}_P = \mathbf{A}_C + \mathbf{A}^n_{P/C} + \mathbf{A}^t_{P/C}$$

(a)   (b)

FIGURE 4.3
Acceleration in rolling contact.

Then by considering the direction of the normal accelerations as being toward the centers of rotation for Fig. 4.3a

$$\mathbf{A}_P = \frac{r_2^2 \omega_2}{r_1 + r_2} - r_2 \omega_2^2$$

$$= -\omega_2^2 \frac{r_1 r_2}{r_1 + r_2}$$

and for Fig. 4.3b

$$\mathbf{A}_P = \frac{r_2^2 \omega_2^2}{r_1 - r_2} + r_2 \omega_2^2$$

$$= \omega_2^2 \frac{r_1 r_2}{r_1 - r_2}$$

Thus the acceleration of the point of contact P (which is an instant center for velocity) is expressed in terms of the radius of curvature of the rolling bodies. If the rolling bodies are polodes, the point P is known as a pole point and the ratio $r_1 r_2 / (r_1 \pm r_2)$ is known as the diameter of an inflection circle.

4.4   CORIOLIS COMPONENT OF ACCELERATION

In the nineteenth century a French mathematician, Coriolis, discovered a component of acceleration associated with a changing radius of rotation. The following intuitive analysis is due principally or in part to Professor G. H. Martin. Figure 4.4 shows link 2 rotating at a constant angular

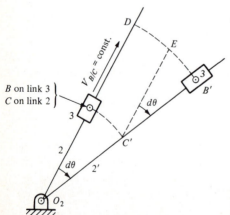

FIGURE 4.4
Coriolis component of acceleration.

velocity ω_2 in a CW sense. Link 3 is a slider which is moving outward along link 2 with a constant velocity with respect to link 2 as link 2 is rotating. Point B is located on link 3, and point C is a coincident point on link 2 at the outset. As link 2 rotates through the small angle $d\theta$, point C on link 2 moves to position C' and point B moves to position B'. The line $C'E$ is parallel to link 2 in its original position. The displacement of point B from B to B' may be expressed as a vector sum of displacements CC', $C'E$, and EB'.

The distance EB' may be calculated in two separate manners both of which must produce the same results.

$$EB' = C'B' \, d\theta \qquad\qquad (a)$$

The distance $C'B'$ will be equal to the velocity $V_{B/C}$ multiplied by the time increment dt, and the angular displacement will be equal to the angular velocity multiplied by the time increment dt. Then

$$C'B' = V_{B/C} \, dt$$

and $$d\theta = \omega_2 \, dt$$

Substitution into Eq. (a) yields

$$EB' = \mathbf{V}_{B/C}\omega_2 \, dt^2 \qquad\qquad (b)$$

The velocity of B directed perpendicular to link 2 is equal to $(O_2B)\omega_2$. Since the angular velocity of link 2 is constant and since link 3 is moving with a constant velocity relative to link 2, the acceleration of B directed perpendicular to link 2 is increasing at a constant rate. The distance traversed by a particle traveling with constant acceleration is equal to one-half the acceleration multiplied by the square of the time increment. Then

$$EB' = \tfrac{1}{2}A \, dt^2 \qquad\qquad (c)$$

Equating Eqs. (b) and (c) and solving for the acceleration of point B directed perpendicular to link 2 yields the Coriolis component of acceleration for point B.

$$\mathbf{A}^c = 2\mathbf{V}_{B/C}\omega_2 \qquad\qquad (4.3)$$

It is important to notice that the Coriolis component of acceleration is directed perpendicular to the vector representing the velocity of sliding of B with respect to C and in a direction indicated by the angular velocity of link 2. In Eq. (4.3) the Coriolis component of acceleration is directed perpendicular to the velocity vector $\mathbf{V}_{B/C}$ in a sense indicated by ω_2.

4.5 GRAPHICAL MECHANISM ANALYSIS

Figure 4.5 shows a Whitworth quick-return mechanism in which the crank, link 1, is moving counterclockwise with angular velocity ω_1 equal to 4 rad/s. The angular velocity of link 1 is increasing at a rate of $\alpha_1 = 5$ rad/s². The angular acceleration of links 3 and 5 and the linear acceleration of link 6 is desired.

The velocity polygon is shown to scale. The normal and tangential accelerations of point B_1 may be computed.

$$\mathbf{A}_{B1}^n = (O_1B_1)\omega_1^2 = 1.0(4^2) = 16.0 \text{ units/s}^2$$

$$\mathbf{A}_{B1}^t = (O_1B_1)\alpha_1 = 1.0(5) = 5.0 \text{ units/s}^2$$

Since the distance O_3B_3 is a changing distance, the Coriolis component of acceleration is involved. Then the acceleration of point B on link 3 may be found from the acceleration of point B on link 2.

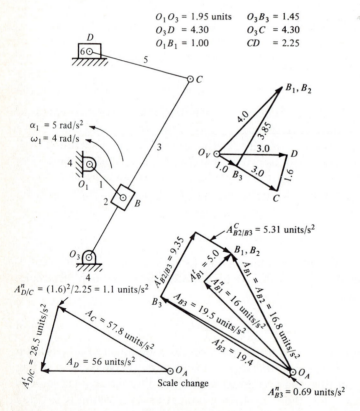

$O_1O_3 = 1.95$ units $O_3B_3 = 1.45$
$O_3D = 4.30$ $O_3C = 4.30$
$O_1B_1 = 1.00$ $CD = 2.25$

$\alpha_1 = 5$ rad/s²
$\omega_1 = 4$ rad/s

$A_{D/C}^n = (1.6)^2/2.25 = 1.1$ units/s²

$A_{B2/B3}^C = 5.31$ units/s²

$A_{B3}^n = 0.69$ units/s²

Scale change

FIGURE 4.5
Whitworth quick-return mechanism.

On link 3, the acceleration of point B is equal to the sum of its normal and tangential accelerations.

$$\mathbf{A}_{B3} = \mathbf{A}_{B3}^{n} + \mathbf{A}_{B3}^{t}$$

$$\mathbf{A}_{B3}^{n} = \frac{V_{B3}^{2}}{O_3 B_3} = \frac{1.0^2}{1.45} = 0.69 \text{ units/s}^2 \text{ directed from } B_3 \text{ toward } O_3$$

$$\mathbf{A}_{B3}^{t} = (O_3 B_3)\alpha_3 \qquad \text{directed perpendicular to } A_{B3}^{n}$$

and
$$\mathbf{A}_{B3} = \mathbf{A}_{B2} + \frac{\mathbf{A}_{B3}^{n}}{B_2} + \frac{\mathbf{A}_{B3}^{t}}{B_2} + \frac{\mathbf{A}_{B3}^{c}}{B_2}$$

In this equation, the acceleration of B on link 2 is known completely. The normal acceleration of B_3 with respect to B_2 is equal to the velocity of B_3 with respect to B_2 squared and divided by the radius of curvature of the path that point B on link 3 will trace on link 2 as the mechanism moves. The normal acceleration of B_3 with respect to B_2 is directed toward the center of curvature of the path that point B on link 3 will trace on link 2. With some effort it will be possible to trace the path of B_3 on link 2 and determine its radius of curvature. However, it is a bit easier to rewrite the equation and solve for the acceleration of B_2 rather than the acceleration of B_3 even though the acceleration of B_2 is already known. Then

$$\mathbf{A}_{B2} = \mathbf{A}_{B3} + \mathbf{A}_{B2/B3}^{n} + \mathbf{A}_{B2/B3}^{t} + \mathbf{A}_{B2/B3}^{c} \qquad (d)$$

Now the normal acceleration of B on link 2 with respect to B on link 3 will be equal to the square of the sliding velocity divided by the radius of curvature of the path that B_2 will trace on link 3 as the mechanism moves. Since link 2 is guided to move along link 3, the radius of curvature of the path that B_2 will trace on link 3 will be infinite since link 3 is straight and the normal acceleration of B on link 2 with respect to B on link 3 will be equal to zero. In the event that link 3 is not a straight line, the normal acceleration of B on link 2 with respect to B on link 3 will be directed toward the center of curvature of link 3.

The tangential acceleration of B on link 2 with respect to B on link 3 will be directed parallel to the relative velocity vector $V_{B2/B3}$, but its sense is unknown.

The Coriolis component of acceleration may be computed from

$$\mathbf{A}_{B2/B3}^{c} = 2V_{B2/B3}\omega_3 = 2(3.85)(0.69) = 5.31 \text{ units/s}^2$$

directed in the direction of $V_{B2/B3}$ rotated 90° in the sense of ω_3. $V_{B2/B3}$ is available from the velocity polygon.

$$\omega_3 = \frac{V_{B3}}{O_3 B_3} = \frac{1.0}{1.45} = 0.69 \text{ rad/s CW}$$

The vectors of Eq. (d) may be added in any order desired to start from the origin of the acceleration polygon and end at the head of the acceleration vector point B_2.

With completion of the acceleration polygon,

$$\alpha_3 = \frac{A^t_{B3}}{O_3 B_3} = \frac{19.4}{1.45} = 13.38 \text{ rad/s}^2 \text{ CCW}$$

$$\alpha_5 = \frac{A^t_{D/C}}{CD} = \frac{28.5}{2.25} = 12.67 \text{ rad/s}^2 \text{ CCW}$$

$$A_{D6} = 56.0 \text{ units/s}^2 \text{ directed to the left}$$

In the cam system of Fig. 4.6, the eccentric (link 2) is rotating counterclockwise with a constant 3 rad/s angular velocity. It is desired to know the acceleration of point B on link 4.

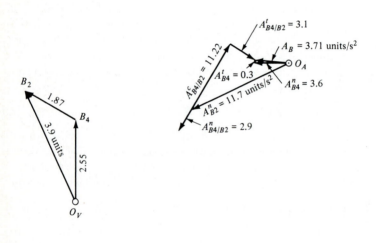

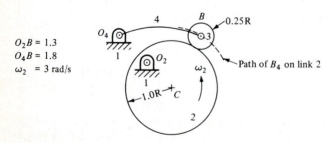

FIGURE 4.6
Cam system acceleration.

In this situation it is important to note that the distance O_2B is a changing distance and that the Coriolis component of acceleration is involved. For problems of this type it is convenient to recognize that three distinctive points exist at point B. There is point B_2 on link 2, point B_3 on link 3, and point B_4 on link 4. Point B_3 is of no interest in this case.

The velocity polygon is constructed in the usual manner with the velocity of point B_2 computed using the distance O_2B_2 and the angular velocity of link 2.

$$\mathbf{V}_{B2} = (O_2B_2)_2 = (1.3)3 = 3.9 \text{ units/s}$$

The velocity of point B_4 is directed perpendicular to O_4B_4 and the velocity of point B_4 with respect to point B_2 is directed tangent to the path that point B_4 will trace on link 2.

The acceleration equation to be solved is

$$\mathbf{A}_{B4} = \mathbf{A}_{B4}^n + \mathbf{A}_{B4}^t = \mathbf{A}_{B2} + \mathbf{A}_{B4/B2}^n + \mathbf{A}_{B4/B2}^t + \mathbf{A}_{B4/B2}^c$$

$$\mathbf{A}_{B4}^n = \frac{(\mathbf{V}_{B4})^2}{O_4B_4} = \frac{(2.55)^2}{1.8} = 3.6 \text{ units/s}^2 \text{ directed from } B_4 \text{ toward } O_4$$

$$\mathbf{A}_{B2} = (O_2B_2)\omega_2^2 = (1.3)(3^2) = 11.7 \text{ units/s}^2 \text{ directed from } B \text{ toward } O_2$$

$$\mathbf{A}_{B4/B2}^n = \frac{\mathbf{V}_{B4/B2}^2}{CB} = \frac{(1.87)^2}{1.25} = 2.8 \text{ units/s}^2 \text{ directed from } B \text{ toward } C$$

$$\mathbf{A}_{B4/B2}^c = 2(\mathbf{V}_{B4/B2})\omega_2 = 2(1.87)3 = 11.22 \text{ units/s}^2 \text{ in the direction of } \mathbf{V}_{B4/B2} \text{ rotated } 90° \text{ CCW}$$

In the completed acceleration polygon, the acceleration of point B on link 4 is found to be 3.71 units/s^2.

4.6 EQUIVALENT MECHANISMS

One of the major problems in acceleration analysis of a point moving with respect to another body is to know the path the point will trace on the moving body. This is particularly true in mechanisms containing sliding or rolling surfaces such as is found in cam systems. Analysis may become quite involved and tedious if a point-to-point technique is attempted. In such instances the equivalent mechanism technique may be of advantage.

Equivalent mechanisms are those which provide motion identical to the mechanism under consideration yet do not involve difficulty in determination of the path traced by a point. The equivalent mechanism may be used for an instantaneous analysis since the equivalent mechanism

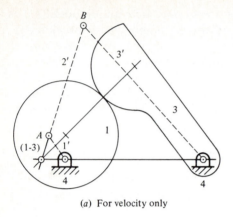

(a) For velocity only

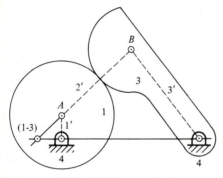

(b) For velocity and acceleration

FIGURE 4.7
Equivalent mechanism.

will most probably change proportions and configuration during a complete cycle of operation.

Some skill is necessary in developing the equivalent mechanism. In several cases it is possible to devise many equivalent mechanisms for the same system. Figure 4.7 shows a set of equivalent mechanisms which can be used for the same original mechanism. Note that there is an infinite number of equivalent mechanisms for this case, yet the two points A and B must lie on a line which passes through the instant center for velocity between the input and output link. Figure 4.8 shows equivalent mechanisms for a series of mechanisms.

4.7 ANALYTIC ACCELERATION ANALYSIS

Acceleration analysis by analytic means requires differentiation of unit vectors. In Fig. 4.9, the unit vector $\mathbf{a}i_r$ at time 1 is located at the angle θ counterclockwise from the x axis and directed along the i_r axis of the moving coordinate system i_r, i_t. At time 2, the unit vector has rotated

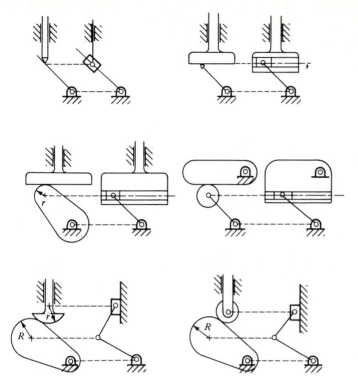

FIGURE 4.8
Equivalent mechanisms.

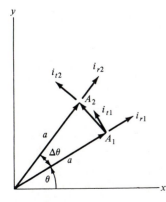

FIGURE 4.9
Differentiation of a unit vector.

through the CCW angle $\Delta\theta$. Change in position of the unit vector terminus is Δa. Then the instantaneous velocity of change is $\Delta\mathbf{ai}_r/\Delta t$ as Δt approaches zero or

$$V_A = \frac{d(\mathbf{ai}_r)}{dt}$$

$$= \frac{da}{dt}\,\mathbf{i}_r + a\,\frac{d\mathbf{i}_r}{dt}$$

Since a unit vector is being considered and its magnitude is constant, da/dt is equal to zero and

$$\mathbf{V}_A = a\,\frac{d\mathbf{i}_r}{dt}$$

From Fig. 4.6,

$$\Delta a = a\,\Delta\theta\,i_t$$

where i_t simply indicates direction.

In the limit as Δt approaches zero

$$\mathbf{V}_A = a\,\frac{d\theta}{dt}\,i_t = a\omega i_t$$

Then

$$\mathbf{V}_A = a\,\frac{d\mathbf{i}_r}{dt} = a\omega i_t$$

or

$$\frac{d\mathbf{i}_r}{dt} = \omega i_t$$

Thus, the differentiation of a unit vector results in another vector of magnitude equal to the angular velocity of the unit vector and directed 90° in the sense of the angular velocity.

Consider now the motion of point P from position 1 to position 2 as shown in Fig. 4.10. Point P is shown as moving in the fixed reference frame x, y. In the moving reference frame r, t has its origin at point P. In the moving reference frame the axis r refers to a radial distance, while the axis t refers to a tangential distance. At time 1, point P is located at P_1 a distance pi_r from the origin of the fixed reference frame. At time 2, point P is located at P_2 a distance $p + \Delta p$ from the origin of the fixed reference frame. The total change in position of point P is then

$$\Delta\mathbf{p} = \mathbf{p}_r + \mathbf{p}_t$$

$$\mathbf{p}_r = \Delta\mathbf{p}\,\mathbf{i}_r$$

$$\mathbf{p}_t = (\mathbf{p} + \Delta\mathbf{p})\,\Delta\theta\,\mathbf{i}_t$$

As Δt approaches zero, $\mathbf{p} + \Delta\mathbf{p}$ approaches $\mathbf{p}$ and in the limit

$$dp = dp\,\mathbf{i}_r + p\,d\theta\,\mathbf{i}_t \tag{e}$$

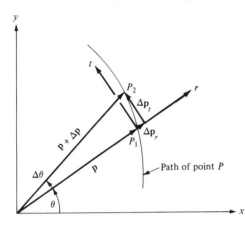

FIGURE 4.10
Vector differentiation.

The velocity of the change in position is the time derivative of equation (e):

$$\mathbf{V}_p = \frac{dp}{dt} = \frac{dp}{dt}\,\mathbf{i}_r + p\,\frac{d\theta}{dt}\,\mathbf{i}_t$$

or

$$\mathbf{V}_p = \dot{p}\mathbf{i}_r + p\omega\mathbf{i}_t \qquad (f)$$

The acceleration associated with the change in position of point P is the time derivative of Eq. (f):

$$\mathbf{A}_p = \frac{dV}{dt} = \ddot{p}\mathbf{i}_r + \dot{p}\,\frac{d\mathbf{i}_r}{dt} + \dot{p}\omega\mathbf{i}_t + p\alpha\mathbf{i}_t + p\omega\,\frac{d\mathbf{i}_t}{dt}$$

In this equation the terms involving differentiation of a unit vector become

$$\dot{p}\,\frac{d\mathbf{i}_r}{dt} = \dot{p}\omega\mathbf{i}_t$$

and

$$p\omega\,\frac{d\mathbf{i}_t}{dt} = -p\omega^2\mathbf{i}_r$$

The minus sign indicates that direction is toward the fixed frame origin.

By collecting terms, the acceleration equation becomes

$$\mathbf{A}_p = (\ddot{p} - p\omega^2)\mathbf{i}_r + (2\dot{p}\omega + p\alpha)\mathbf{i}_t$$

where $\ddot{p}\mathbf{i}_r$ = change in length of position vector directed radially outward or inward depending on increasing or decreasing length
$-p\omega^2\mathbf{i}_r$ = usual normal acceleration directed toward center of curvature of path of P
$2\dot{p}\omega\mathbf{i}_t$ = Coriolis component of acceleration
$p\alpha\mathbf{i}_t$ = usual tangential acceleration directed as indicated by sense of α

4.8   ACCELERATION ANALYSIS WITH POLAR COMPLEX NOTATION

With the position vector expressed in polar complex notation

$$\mathbf{P} = pe^{i\theta}$$

Differentiation with respect to time provides the velocity.

$$\mathbf{V}_p = \frac{dp}{dt}\, e^{i\theta} + ipe^{i\theta}\, \frac{d\theta}{dt}$$

$$= \dot{p}e^{i\theta} + ip\omega e^{i\theta}$$

A second differentiation with respect to time provides the acceleration.

$$\mathbf{A}_p = \frac{d^2p}{dt^2}\, e^{i\theta} + i\, \frac{dp}{dt}\, e^{i\theta}\, \frac{d\theta}{dt}$$

$$+ i\, \frac{dp}{dt}\, e^{i\theta}\, \frac{d\theta}{dt} + i^2 pe^{i\theta}\left(\frac{d\theta}{dt}\right)^2$$

$$+ ipe^{i\theta}\, \frac{d^2\theta}{dt^2}$$

or
$$\mathbf{A}_p = \ddot{p}e^{i\theta} + 2i\dot{p}\omega e^{i\theta} - p\omega^2 e^{i\theta} + ip\alpha e^{i\theta}$$

where $\ddot{p}e^{i\theta}$ = acceleration with which magnitude of vector is changing

$2i\dot{p}\omega e^{i\theta}$ = Coriolis component of the acceleration; i indicates a 90° rotation in direction of θ

$p\omega^2 e^{i\theta}$ = normal acceleration; minus sign indicates that normal acceleration is directed toward origin of vector

$ip\alpha e^{i\theta}$ = tangential acceleration; i indicates a 90° rotation in direction of θ

4.9   ACCELERATION ANALYSIS USING COMPLEX NUMBERS

With the links of a four-bar mechanism (Fig. 4.11) represented by vectors using polar complex numbers, the distance $O_A A$ to B may be computed using two separate paths. This is known as the configuration equation.

$$r_1 e^{i\theta_1} + r_2 e^{i\theta_2} = r_4 + r_3 e^{i\theta_3}$$

Differentiation of the configuration equation with respect to time will provide the velocity equation in which ω represents angular velocity.

$$r_1\omega_1 e^{i\theta_1} + r_2\omega_2 e^{i\theta_2} = r_3\omega_3 e^{i\theta_3}$$

Differentiation of the velocity equation with respect to time will provide the acceleration equation in which α represents angular acceleration.

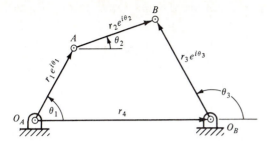

FIGURE 4.11
Acceleration analysis with complex numbers.

$$r_1\alpha_1 e^{i\theta_1} + ir_1\omega_1^2 e^{i\theta_1} + r_2\alpha_2 e^{i\theta_2} + ir_2\omega_2^2 e^{i\epsilon\theta_2} = r_3\alpha_3 e^{i\theta_3} + ir_3\omega_3^2 e^{i\theta_3}$$

Expanding the acceleration equation in terms of the sin and cos functions provides

$$r_1\alpha_1(\cos\theta_1 + i\sin\theta_1) + ir_1\omega_1^2(-\sin\theta_1 + i\cos\theta_1)$$
$$+ r_2\alpha_2(\cos\theta_2 + i\sin\theta_2) + ir_2\omega_2^2(-\sin\theta_2 + i\cos\theta_2)$$
$$= r_3\alpha_3(\cos\theta_3 + i\sin\theta_3) + ir_3\omega_3^2(-\sin\theta_3 + i\cos\theta_3)$$

Separating the real and the imaginary parts results in two equations,

$$r_1\alpha_1\cos\theta_1 - r_1\omega_1^2\sin\theta_1 + r_2\alpha_2\cos\theta_2 - r_2\omega_2^2\sin\theta_2$$
$$= r_3\alpha_3\cos\theta_3 - r_3\omega_3^2\sin\theta_3$$
$$r_1\alpha_1\sin\theta_1 - r_1\omega_1^2\cos\theta_1 + r_2\alpha_2\sin\theta_2 - r_2\omega_2^2\cos\theta_2$$
$$= r_3\alpha_3\sin\theta_3 - r_3\omega_3^2\cos\theta_3$$

In these two equations, the known quantities are r_1, r_2, r_3, r_4, θ_1, ω_1, and α_1. The unknown quantities are ω_2, ω_3, α_2, α_3, θ_2, and θ_3.

However, θ_2 and θ_3 are available from Eqs. (2.2) and ω_2 and ω_3 are available from Eqs. (3.2) and (3.3). The only remaining unknowns are α_2 and α_3.

The equations are of the form:

$$A\alpha_2 + B\alpha_3 = C$$
$$F\alpha_2 + G\alpha_3 = H$$

where $A = r_2\cos\theta_2$
$B = -r_3\cos\theta_3$
$C = r_1\omega_1^2\sin\theta_1 - r_1\alpha_1\cos\theta_1 + r_2\omega_2^2\sin\theta_2 - r_3\omega_3^2\sin\theta_3$
$F = r_2\sin\theta_2$
$G = -r_3\sin\theta_3$
$H = r_3\omega_3^2\cos\theta_3 - r_1\alpha_1\sin\theta_1 - r_1\omega_1^2\cos\theta_1 - r_2^2\omega_2^2\cos\theta_2$

The two equations may be solved for the angular acceleration of the coupler and rocker links using Cramer's rule.

$$\alpha_2 = \frac{\begin{vmatrix} C & B \\ H & G \end{vmatrix}}{\begin{vmatrix} A & B \\ F & G \end{vmatrix}} = \frac{CG - HB}{AG - FB}$$

$$\alpha_3 = \frac{\begin{vmatrix} A & C \\ F & H \end{vmatrix}}{\begin{vmatrix} A & B \\ F & G \end{vmatrix}} = \frac{AH - FC}{AG - FB}$$

4.10 COMPUTER PROGRAM FOR ACCELERATION ANALYSIS

The equations for α_2 and α_3 [together with Eqs. (2.2), (3.2), and (3.3)] may be solved using the computer program ACCELERATION (App. A).

Output from program ACCELERATION will be

THIS PROGRAM PROVIDES ANGULAR ACCELERATION OF EACH LINK OF A FOUR-BAR MECHANISM (COUNTERCLOCKWISE POSITIVE) OPEN OR CROSSED MECHANISMS ARE CONSIDERED. VELOCITIES ARE IN RADIANS PER SECOND. ACCELERATIONS IN RADIANS PER SECOND SQUARED.

```
LENGTH OF FIXED LINK          = 10
LENGTH OF CRANK               = 4
LENGTH OF COUPLER             = 12
LENGTH OF ROCKER              = 8
CRANK ANGULAR VELOCITY        = 2
CRANK ANGULAR ACCELERATION    = 2
```

MECHANISM IS OPEN

CRANK ANGLE	COUPLER ANGULAR ACCELERATION	ROCKER ANGULAR ACCELERATION
0	0.102463	−1.859311
45	−1.208279	2.821716
90	2.141205	0.082346
135	1.156276	0.632570
180	1.037497	0.411751
225	−2.059630	−2.923578
270	−2.046580	−1.154043
315	0.099061	1.298803
360	0.102463	−1.859311

4.11 COUPLER POINT ACCELERATION

The acceleration of a point on the coupler link may be found with reference to Fig. 4.12. The distance O_AC may be found as $O_AC = O_AA + AC$. In polar vector notation, the configuration equation becomes

$$O_AC = r_1 e^{i\theta_1} + RC e^{i(\theta_2 + \epsilon)}$$

The velocity of point C is found by differentiating the configuration equation remembering that ϵ is a constant.

$$\mathbf{V}_C = r_1 \omega_1 e^{i\theta_1} + RC\omega_2 e^{i(\theta_2 + \epsilon)}$$

Acceleration of point C is found by differentiating the velocity equation.

$$\mathbf{A}_C = r_1 \alpha_1 e^{i\theta_1} + ir_1 \omega_1^2 e^{i\theta_1} + RC\alpha_2 e^{i(\theta_2 + \epsilon)} + iRC\omega_2^2 e^{i(\theta_2 + \epsilon)}$$

Expanding the acceleration equation in terms of sin and cos functions gives

$$\begin{aligned}
\mathbf{A}_C = {}& r_1 \alpha_1 (\cos \theta_1 + i \sin \theta_1) + ir_1 \omega_1^2 (\cos \theta_1 + i \sin \theta_1) \\
& + RC\alpha_2 [\cos (\theta_2 + \epsilon) + i \sin (\theta_2 + \epsilon)] \\
& + iRC\omega_2^2 [\cos (\theta_2 + \epsilon) + i \sin (\theta_2 + \epsilon)]
\end{aligned}$$

Separating the real and the imaginary components will provide the x and y components of the acceleration of point C.

$$AC_x = r_1 \alpha_1 \cos \theta_1 - r_1 \omega_1^2 \sin \theta_1 + RC\alpha_2 \cos (\theta_2 + \epsilon) - RC\omega_2^2 \sin (\theta_2 + \epsilon)$$

$$AC_y = r_1 \alpha_1 \sin \theta_1 + r_1 \omega_1^2 \cos \theta_1 + RC\alpha_2 \sin (\theta_2 + \epsilon) + RC\omega_2^2 \cos (\theta_2 + \epsilon)$$

The magnitude of acceleration is the square root of the sum of the squares of the x and y components of acceleration. However, for later use in computing acceleration forces it is best to have the x and y components of acceleration.

$$\mathbf{A}_C = \mathbf{A}_{Cx}^2 + \mathbf{A}_{Cy}^2$$

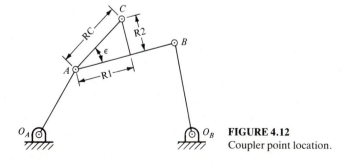

FIGURE 4.12
Coupler point location.

The direction of the acceleration is given by

$$\text{Direction } \mathbf{A}_C = \arctan \frac{\mathbf{A}_{Cy}}{\mathbf{A}_{Cx}}$$

Care must be taken to recognize the quadrant in which the direction of acceleration exists. In order to use these equations it is necessary to evaluate ϵ properly as shown in Fig. 4.13 and lines 271, 272, 273, 274 of the following modified program.

The program for angular acceleration of links may be modified to provide the acceleration of a coupler point giving program CPACCELE-RATION (App. A).

Output from program CPACCELERATION will be

* * * * * * * * * * COUPLER POINT ACCELERATION * * * * * * * * * *

THIS PROGRAM PROVIDES THE ACCELERATION OF A COUPLER POINT OF AN OPEN OR CROSSED FOUR-BAR MECHANISM. CRANK ANGLE IN DEGREES. VELOCITIES IN RADIANS PER SECOND. ACCELERATIONS IN RADIANS PER SECOND SQUARED.

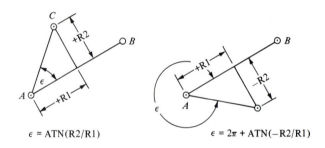

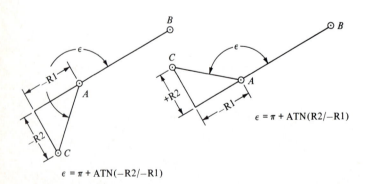

FIGURE 4.13
Location of coupler point.

LENGTH OF FIXED LINK = 10
LENGTH OF CRANK = 4
LENGTH OF COUPLER = 12
LENGTH OF ROCKER = 8

DISTANCE FROM CRANK PIN TO PERPENDICULAR TO COUPLER POINT   6
PERPENDICULAR DISTANCE FROM COUPLER TO COUPLER POINT   3

CRANK ANGULAR VELOCITY = 2
CRANK ANGULAR ACCELERATION = 2

MECHANISM IS OPEN

CRANK	COUPLER POINT	
ANGLE	X ACCEL.	Y ACCEL.
0	2.180218	30.80655
30	7.558782	30.58688
60	−5.121197	19.71297
90	−12.548250	11.66482
120	−14.577170	3.42675
150	−12.351620	−4.15710
180	−7.445667	−10.41876
210	−1.021575	−14.83260
240	6.605785	−17.56341
270	14.607720	−19.55309
300	18.507460	−23.77799
330	3.499324	−20.04411
360	2.180218	30.80655

PROBLEMS

4.1. Link 1 has an angular velocity of 15 rad/s CCW which is increasing at the rate of 75 rad/s². Determine the acceleration of point A.

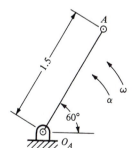

FIGURE P4.1
Crank point acceleration.

4.2. An automobile with 28-in outside diameter tires is moving at a rate of 45 mi/h and increasing speed at the rate of 30 mi/h each minute. Determine the acceleration of a point on the top of the tire.

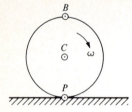

FIGURE P4.2
Rolling acceleration.

4.3. For the slider crank mechanism with $\omega_1 = 10 \, \text{rad/s}$ and $\alpha_1 = 50 \, \text{rad/s}^2$ determine the linear acceleration of points A and C and the angular velocity and acceleration of link 2.

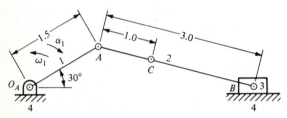

FIGURE P4.3
Slider crank.

4.4. The angular velocity of the driving link is $\omega_1 = 4.0 \, \text{rad/s}$ CCW. The angular velocity is increasing at the rate of $10 \, \text{rad/s}^2$. Determine the linear acceleration of points B and C and the angular velocity and acceleration of links 2 and 3.

$$O_AO_B = 1.75 \qquad O_AA = 0.75 \qquad O_BB = 1.50$$
$$AB = 2.00 \qquad AC = 1.50 \qquad BC = 1.50$$

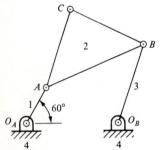

FIGURE P4.4
Coupler link acceleration.

4.5. With the angular velocity of link 1 equal to $5.0 \, \text{rad/s}$ CCW and its angular acceleration $25 \, \text{rad/s}^2$ CCW determine the linear acceleration of points G_1, G_2, and B and the angular acceleration of link 2.

$$O_AA = 1.5 \qquad O_AG_1 = 1.0 \qquad AB = 3.5 \qquad AG_2 = 1.25$$

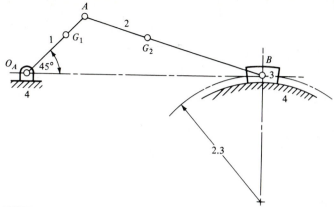

FIGURE P4.5
Centers of gravity acceleration.

4.6. Determine the linear acceleration of point C, and find the angular acceleration of links 2 and 3. Locate the instant center for acceleration (Σ) for
link 2.

$$O_A A = 1.5 \qquad O_A O_B = 3.00 \qquad O_B B = 2.0$$
$$AC = 2.0 \qquad BC = 1.5$$
$$\omega_1 = 3 \, \text{rad/s CCW} \qquad \alpha_1 = 8 \, \text{rad/s}^2 \, \text{CCW}$$

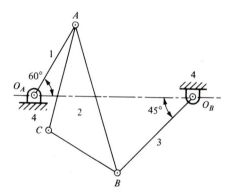

FIGURE P4.6
Acceleration analysis.

4.7. Plot the path of point C of the Dobbie-McGinnis engine indicator, and
determine the velocity ratio V_C/V_B. If ω_1 is constant at $1.0 \, \text{rad/s}$, determine
the acceleration ratio A_C/A_B.

$$FC = 5.10 \qquad FE = 1.60 \qquad BD = 1.10$$
$$O_E E = 0.90 \qquad O_E D = 0.40$$

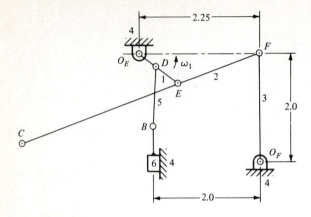

FIGURE P4.7
Engine indicator acceleration.

4.8. Construct an acceleration polygon, and determine the linear acceleration of points C and D. Find the angular acceleration of links 3 and 5.

$$O_B B = 0.5 \qquad O_C O_B = 2.0 \qquad CD = 1.5$$
$$\omega_1 = 3.0 \, \text{rad/s CCW} \qquad\qquad \alpha_1 = 2.0 \, \text{rad/s}^2 \, \text{CCW}$$

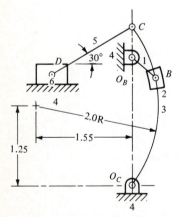

FIGURE P4.8
Acceleration polygon.

4.9. Determine the linear acceleration of point D on link 5 and the angular acceleration of link 3.

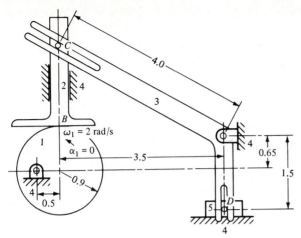

FIGURE P4.9.
Acceleration analysis.

4.10. Link 3 rolls on link 4 without slipping. Determine the angular acceleration of link 5.

$$O_B B = 1.25 \qquad BC = 2.50 \qquad DE = 2.50 \qquad CD = 0.75$$
$$\omega_1 = 4.0 \, \text{rad/s} \qquad\qquad \alpha_1 = 0 \, \text{rad/s}^2$$

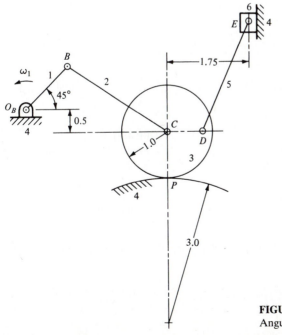

FIGURE P4.10
Angular acceleration.

4.11. With the angular velocity of link 1 equal to 2 rad/s CCW and increasing at the rate of 4 rad/s^2 determine

(a) The angular acceleration of links 2 and 3.
(b) The acceleration of point C.

Check the results using a computer program.

$$O_A O_B = 2.5 \qquad O_A A = 1.0 \qquad O_B B = 2.0 \qquad AB = 4.0$$
$$AD = 2.0 \qquad DC = 0.75$$

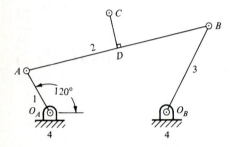

FIGURE P4.11
Acceleration analysis.

4.12. Acceleration of the endpoints of the link are given. Determine the angular acceleration of the link.

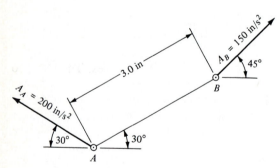

FIGURE P4.12
Floating link acceleration.

CHAPTER
5

STATIC
AND
INERTIA
FORCES

5.1 INTRODUCTION

Forces involved in mechanisms result from several sources. The loads to be overcome in doing useful work are generally related to static forces such as might be encountered in heavy, slow-moving machinery. In the case of high-speed machinery the loads due to inertia forces are apt to result in the highest forces on the machine members. In some cases the shear weight of the machine component may contribute significantly to the force in the members.

Force analysis (whether it be static or dynamic) relies on the principles of statics in that the sum of forces in any direction and the sum of moments about any point must be zero. The forces are considered as vectors and thus have magnitude as well as direction. The line of action of a force is along the vector representing the force.

Particular attention must be paid to the subscripts describing a force. A force indicated with only one subscript represents an external force acting on the link designated by the subscript. A force indicated with two subscripts represents a force of one link (the first subscript) acting on another link (the second subscript). Thus $\mathbf{F}_{32}$ is the force that link 3 will exert on link 2 and is directed through the connection between links 2 and 3. In like manner, $\mathbf{F}_{23}$ is the force that link 2 will exert on link 3, is equal and opposite to $\mathbf{F}_{32}$, and has the same line of action.

5.2 STATIC FORCES

In general, a link of a mechanism which is connected with revolute joints to two other links is a two-force system. As shown in Fig. 5.1, since link 2 is free to rotate about the center of its revolute joints, a force acting on a two-force link (in the absence of friction) must act along the centerline of that link. The direction of the resultant force acting on a tertiary, or three-connection, link must be found by summation of moments or forces.

Since the forces are considered as vectors, all the rules, regulations, and techniques associated with vectors may be used. In Fig. 5.2, the force $\mathbf{F}_3$ is an external force acting on link 3 of the mechanism. A complete force analysis will provide knowledge of all forces acting at revolute joints as well as of the torque which must be applied to link 1 to maintain static equilibrium of the system.

Starting with link 3, there are three forces acting: $\mathbf{F}_3$, $\mathbf{F}_{23}$, and $\mathbf{F}_{43}$. Since these three forces constitute a state of static equilibrium for link 3, the sum of the three forces must be zero. The three forces acting on link 3 comprise a coplanar concurrent force system and as such all lines of action of the forces must intersect at a common point. Link 2 is a two-force system, and the line of action of forces acting on link 2 is along the centerline of the link. Extending the centerline of link 2 until it intersects with $\mathbf{F}_3$ provides a point through which the line of action of $\mathbf{F}_{43}$ must pass. Since $\mathbf{F}_{43}$ must also pass through point O_B, the line of action of $\mathbf{F}_{43}$ is known. With the lines of $\mathbf{F}_{23}$ and $\mathbf{F}_{43}$ both known and the direction and magnitude of $\mathbf{F}_3$ known, a force polygon will provide the magnitude of $\mathbf{F}_{23}$ and $\mathbf{F}_{43}$.

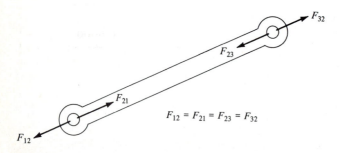

$$F_{12} = F_{21} = F_{23} = F_{32}$$

FIGURE 5.1
Link as a two-force system.

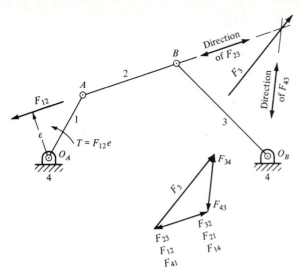

FIGURE 5.2
Static force analysis

The force $\mathbf{F}_{32}$ is equal and opposite to the force $\mathbf{F}_{23}$ and, since $\mathbf{F}_{32}$ and $\mathbf{F}_{12}$ are the only two forces acting on link 2, the force $\mathbf{F}_{12}$ is equal and opposite in direction to the force $\mathbf{F}_{32}$.

In order for link 1 to be a driver of the system, it must be able to receive a torque through point O_A. Acting on link 1 then is the two forces $\mathbf{F}_{21}$ and $\mathbf{F}_{41}$ as well as a torque $\mathbf{T}_1$. The force $\mathbf{F}_{21}$ must be equal and opposite to the force $\mathbf{F}_{12}$ and must pass through the point A. With only two forces acting on link 1, the force $\mathbf{F}_{41}$ must be equal and opposite to the force $\mathbf{F}_{21}$ and must also act through point O_A. The applied torque $\mathbf{T}_1$, through O_A, will be equal and opposite to the torque resulting from force $\mathbf{F}_{21}$ acting about point O_A.

The automotive hood hinge assembly is shown in Fig. 5.3 with the hood half open. The weight of the hood may be considered as a static load applied through the center of gravity of link 2. The spring is intended to completely counterbalance the weight of the hood when the hood is half open. When the hood is completely open, the spring will more than compensate for the hood weight. When the hood is closed, the spring will undercompensate for the hood weight.

With the mechanism half open, it is desired to determine the force which the spring must exert on link 1 to completely counterbalance the hood weight. Link 2 may be considered as a free body with three forces acting on it: $\mathbf{F}_2$, $\mathbf{F}_{32}$, and $\mathbf{F}_{12}$. The lines of action of the three acting forces must intersect at one point. Force $\mathbf{F}_{32}$ must act along the axis of link 3 since link 3 is a two-force system. Force $\mathbf{F}_{12}$ must act through point A,

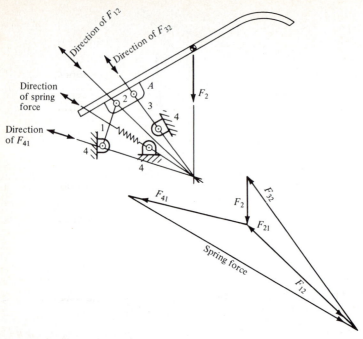

FIGURE 5.3
Static force analysis of an auto hood hinge mechanism.

and its line of action must pass through the intersection of the lines of action of forces F_2 and F_{32}. A force polygon gives the magnitude of forces F_{12} and F_{32}.

Link 1 has three forces acting on it: F_{21}, F_{41}, and the spring force. Force F_{21} is equal and opposite to force F_{12}. The lines of action of the three forces must intersect at a common point, and a force polygon gives their magnitude.

5.3 FRICTION

In machines, friction can be either good or bad. In the case of brakes, friction of the brake shoes on the drum or disk will allow a load to be stopped. In clutches, friction is used to gently engage machine components and then to lock the system to avoid slip. In mechanisms, friction generally acts to alter the direction of the link forces. Fortunately, the alteration is not large, and for this reason the effect of friction is often neglected.

The force due to friction is given by the product of the coefficient of friction and the force normal to the surface where the friction force is acting. The coefficient of friction for components in contact which are at rest and starting to move is generally higher than the coefficient of friction for steady-state motion. It is interesting to note that the friction force is a product of only the normal force and the coefficient of friction. The area of contact surface does not enter into the calculations.

In mechanisms, friction forces exist in one of two types, both of which act to oppose motion. In the case of the slider crank mechanism there is sliding friction between the slider (piston) and the frame (cylinder). Between the crank and the coupler (connecting rod) a rotating pin friction force exists. Although they are identical in nature, the two types of friction force are treated a little differently.

Sliding friction is illustrated in Fig. 5.4. The normal force is $\mathbf{F}_{43}$ and the coefficient of friction is μ. With the inclusion of the friction force in the analysis, the force $\mathbf{F}_{43}$ is no longer directed perpendicular to the direction of motion. The force is inclined at an angle ϕ from the normal.

$$\phi = \arctan \frac{\mu \mathbf{F}_{43}}{\mathbf{F}_{43}} = \arctan \mu$$

Rotating pin friction is illustrated in Fig. 5.5a. The pin is firmly attached to link 2 and is free to move in link 3, and the force is being transmitted from link 2 to link 3. An expanded view of the pin connection is shown in Fig. 5.5b. The force $\mathbf{F}$ creates a friction force $\mu \mathbf{F}$ directed so as to oppose the relative motion. The resultant of the applied force $\mathbf{F}$ and the friction force $\mu \mathbf{F}$ is equal to the force which link 2 applies on link 3, or $\mathbf{F}_{23}$.

$$\mathbf{F}_{23} = \sqrt{\mathbf{F}^2 + (\mu \mathbf{F})^2} = \mathbf{F}\sqrt{1 + \mu^2} = \mathbf{F}\sqrt{1 + \tan^2 \phi}$$
$$= \mathbf{F} \sec \phi$$

The moment of the resultant force about the pin center must be equal to the moment of the friction force about the pin center. Therefore,

$$\mathbf{F}_{23}r = \mu \mathbf{F}R = (\mathbf{F} \sec \phi)r$$

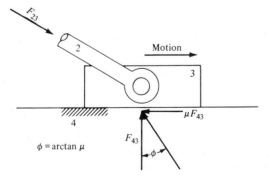

$\phi = \arctan \mu$

FIGURE 5.4
Sliding friction.

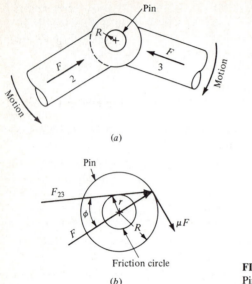

(a)

Pin

F_{23}

Friction circle

FIGURE 5.5
Pin friction.

(b)

Solving for r and remembering that $\mu = \tan \phi$ gives

$$r = R \sin \phi$$

The "friction circle" has radius r, and the resultant force must pass tangent to the friction circle. For a pin of 1.0 in diameter and a coefficient of friction equal to 0.15, the radius of the friction circle becomes 0.148 in or a little over $\frac{1}{8}$ in.

5.4 POLE FORCE ANALYSIS

The pole force method of analysis, which was introduced by Hain,[*] provides a simple method for force analysis of a mechanism. The technique requires use of an auxiliary line which in many cases can be an arbitrary line. However, if the forces involved result from such things as springs connected between two links, the auxiliary line must be a collineation axis. In Fig. 5.6 the three diagonal collineation axes are shown as $L(13, 24)$, $L(14, 23)$, and $L(12, 34)$.

The mechanism of Fig. 5.7 is acted upon by two external forces. One completely known force acting on link 3 tends to cause CW rotation of link 1. Since the force comes from outside the mechanism, it is

---

[*] K. Hain, *Applied Kinematics*, McGraw-Hill, New York, 1967 (trans. from *Angewordte Getriebelehre*, VDI-Verlag GmbH, Düsseldorf, 1961).

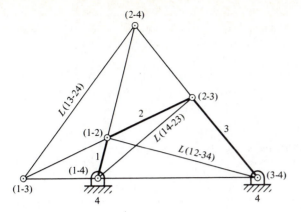

FIGURE 5.6
Collineation axes.

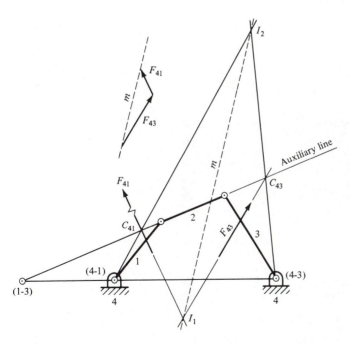

FIGURE 5.7
Pole force analysis.

designated as $\mathbf{F}_{43}$. The other unknown force acting on link 1 from an outside source and designated as $\mathbf{F}_{41}$ would tend to rotate link 1 in a CCW direction. It is desired to determine the magnitude of the force $\mathbf{F}_{41}$ that would place the system in equilibrium.

Point I_1 is located at the intersection of the lines of action of the two forces $\mathbf{F}_{43}$ and $\mathbf{F}_{41}$. Since the two forces are acting on links 1 and 3, the instant center for velocity $(1, 3)$ must be used. Through center $(1, 3)$ an auxiliary line is drawn. The line may be in any arbitrary direction, but it is convenient to use an auxiliary line which is coincident with link 2. The lines of action of the two forces are extended to intersect the auxiliary line at points C_{41} and C_{43}. From C_{41} a line is drawn through instant center $(4, 1)$ and extended to intersect a line through C_{43} and the instant center $(4, 3)$ at point I_2. The line m connects points I_1 and I_2. In the force polygon, the direction of the unknown force $\mathbf{F}_{41}$ is added to the known force $\mathbf{F}_{43}$ and extended to intersect a line through the origin of the force polygon and parallel to the line m. In this manner the magnitude of the unknown force $\mathbf{F}_{41}$ may be determined.

For proof of the procedure (Fig. 5.8) the force $\mathbf{F}_{43}$ is replaced by two components. One component $\mathbf{f}_{43}$ is along the auxiliary line, and the other is along the line connecting C_{43} and instant center $(4, 3)$. The force $\mathbf{F}_{41}$ is replaced by two components. One component $\mathbf{f}_{41}$ is along the

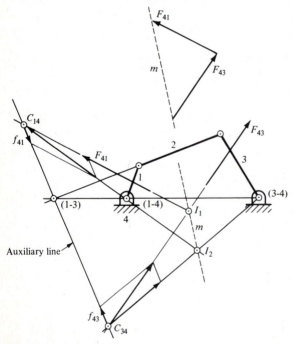

FIGURE 5.8
Proof of pole force procedure.

auxiliary line and the other is along the line connecting C_{41} and instant center $(4, 1)$.

The instant center $(1, 3)$ consists of two coincident points, one on link 1 and the other on link 3. As such, the two points must have the same velocity. By the principle of virtual work then the force $\mathbf{f}_{43}$ must be equal in magnitude and opposite in direction to the force $\mathbf{f}_{41}$. Thus the two force components will cancel one another leaving only the two components whose lines of action pass through point I_2.

In the case of a spring force between two links as in Fig. 5.9, selection of the auxiliary line is not arbitrary. The external force acting is $\mathbf{F}_{43}$, and it is desired to locate the required spring force to maintain equilibrium. The spring force may be considered as $\mathbf{F}_{12}$ or as $\mathbf{F}_{21}$. If the spring force is taken as $\mathbf{F}_{12}$ and the external force is $\mathbf{F}_{43}$, the auxiliary line must be the collineation axis $L(14, 23)$. The line is designated by using the first two and the last two subscripts of the force designations. If the forces acting are considered as $\mathbf{F}_{21}$ and $\mathbf{F}_{43}$, the auxiliary line must be the collineation axis $L(24, 13)$.

Point I_1 is located at the intersection of the lines of action of the two forces considered. Using the two forces as $\mathbf{F}_{12}$ and $\mathbf{F}_{43}$, the points C_{12} and C_{43} are located at the intersection of the lines of action of the forces and the collineation axis $L(14, 23)$. Point I_2 is located at the intersection of the line connecting C_{12} and instant center $(1, 2)$ with the line connecting C_{43} and the instant center $(4, 3)$. Line m passes through I_1 and I_2.

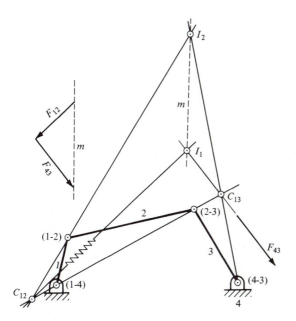

FIGURE 5.9
Pole force procedure with a spring force.

5.5 CONCEPTS OF INERTIA FORCES

In accordance with Newton's laws of motion, if a movable mass is acted upon by a force, an acceleration of the mass results. If the force is designated as $\mathbf{F}$, the mass as M, and the velocity of the mass by $\mathbf{V}$, then Newton's laws of motion provide

$$\mathbf{F} = \frac{d(M\mathbf{V})}{dt} = M\frac{d\mathbf{V}}{dt} + \mathbf{V}\frac{dM}{dt}$$

Since the mass of machine parts generally does not change during motion of the machine, the term $\mathbf{V}\,dM/dt$ is zero. In the case of changing mass (as may be encountered in the reduced fuel mass in a rocket), the changing mass term is not zero. For machine components, Newton's laws of motion provide

$$\mathbf{F} = M\frac{d\mathbf{V}}{dt} = M\mathbf{A}$$

where $\mathbf{F}$ = inertia force acting in direction of acceleration

M = mass of machine member

$\mathbf{A}$ = acceleration of machine member

D'Alembert's principle is a corollary of Newton's laws and states that, if a mass is moving with an acceleration, a force results. That force is known as an inertia force. In kinematics, d'Alembert's principle is used and interest is centered on the force resulting from an acceleration rather than on the force causing an acceleration as expressed in Newton's laws. Therefore, the inertia force is always directed in a sense opposite that of the acceleration. Application of d'Alembert's principle results in a state of kinetostatic equilibrium and all the laws of statics may be applied.

Figure 5.10a shows a body of total mass M moving with angular velocity $\boldsymbol{\omega}$ and angular acceleration $\boldsymbol{\alpha}$ so that the center of gravity of the body has an acceleration of $\mathbf{A}_G$. Interest is centered on arbitrary point B which has a mass of dM and an acceleration which will be equal to the sum of the acceleration of the center of mass of the body plus the normal and tangential accelerations of the point B with respect to the center of gravity as shown in Fig. 5.10a. The total inertia force acting at point B may be considered to be the sum of three inertia forces all directed opposite in sense to the related acceleration. The point B is an arbitrary point with mass dM. The sum of the mass of all points like B in the body must be the total mass of the body. In like manner, the sum of all the forces acting on the differential mass of arbitrary points, like point B, will be the total inertia force acting on the body.

Considering such forces as $dM\,\mathbf{A}_G$,

$$\mathbf{F}_1 = \sum dM\,\mathbf{A}_G = M\mathbf{A}_G$$

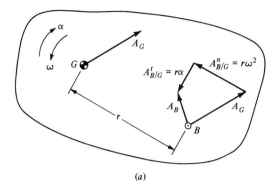

(a)

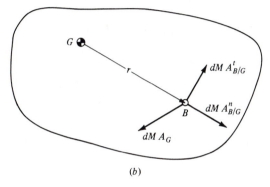

(b)

FIGURE 5.10
Inertia forces due to acceleration.

All forces like $\mathbf{F}_1$ will be directed parallel to the acceleration of the center of gravity and opposite in sense. Since the little mass units dM are all distributed within the body about the center of gravity, the resultant force $\mathbf{F}_1$ will pass through the center of gravity parallel to and opposite in sense of the acceleration of the center of gravity.

Considering such forces as $dM \, \mathbf{A}_{B/G}^n$,

$$\mathbf{F}_2 = \sum dM \, \mathbf{A}_{B/G}^n = \omega^2 \sum r \, dM = 0$$

All forces like $\mathbf{F}_2$ will be directed through the center of gravity of the body. The term $\sum r \, dM$ defines the location of the center of gravity of a body, and since r is in this case measured from the center of gravity, the net result of all forces like $\mathbf{F}_2$ will be zero.

Considering such forces as $dM \, \mathbf{A}_{B/G}^t$,

$$\mathbf{F}_3 = \sum dM \, \mathbf{A}_{B/G}^t = \alpha \sum r \, dM$$

Again the term $\sum r \, dM$ is encountered. However, in this case the direction of the inertia force $\mathbf{F}_3$ is perpendicular to the radius to the center of gravity.

All the $\mathbf{F}_3$-type forces are not parallel to one another nor do they pass through the same point. Therefore, an inertial torque results. The torque will be

$$\mathbf{T} = \sum r\,\mathbf{F}_3 = \alpha \sum r^2\,dM = \alpha I$$

The term $\sum r^2\,dM$ is recognized as an expression for the mass moment of inertia of the body.

In summary, acceleration of the body results in a force of magnitude $\mathbf{F}_1 = MA_G$ which passes through the center of gravity of the body in a direction opposite to the acceleration of the center of gravity and a torque of magnitude $\mathbf{T} = I\alpha$ operating on the body in a sense opposite to the angular acceleration of the body. The total inertia forces are then as shown in Fig. 5.11.

Since a torque may be moved about its plane of action without changing any of the body action, the torque is moved to align one of the forces of the torque parallel to and opposite to the force $\mathbf{F}_1$. One of the forces is made equal to the force $\mathbf{F}_1$, and the two forces are separated by a distance h.

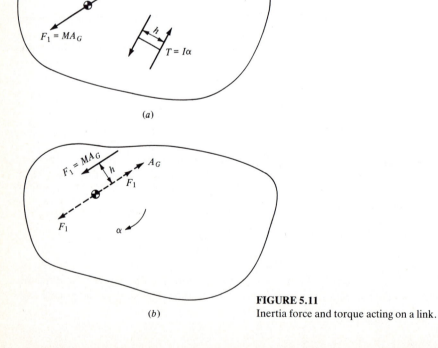

(a)

(b)

FIGURE 5.11
Inertia force and torque acting on a link.

In this manner, the total resultant inertia influence on the body will be a force of magnitude MA_G directed opposite in sense to the acceleration of the center of gravity and offset from the center of gravity an amount h such that the resultant torque about the center of gravity will be in a sense opposite to the angular acceleration of the body as shown in Fig. 5.11b.

The link of Fig. 5.12a has a mass M and a mass moment of inertia I and is rotating about a fixed center with an angular velocity of ω and angular acceleration α. The center of gravity of the link (located at point G) has normal and tangential acceleration resulting in total acceleration $\mathbf{A}_G$ as shown. The total inertia force is $\mathbf{F} = MA_G$ located a distance $h = I\alpha/F$ from the center of gravity such that the torque created by $\mathbf{T} = \mathbf{F}h$ is in a sense opposite to the angular acceleration of the link, α, as shown in Fig. 5.12b. The two triangles GCB and GDE are similar. Therefore,

$$\frac{GC}{GD} = \frac{GB}{GE}$$

or
$$GB = \frac{(GC)(GE)}{GD}$$

but
$$GC = \frac{I\alpha}{MA_G}$$

$$GE = \mathbf{A}_G$$

$$GD = \mathbf{A}_G^t = (OG)\alpha$$

Then
$$GB = \frac{(I\alpha/MA_G)\mathbf{A}_G}{(OG)\alpha} = \frac{I}{(OG)M} \tag{5.1}$$

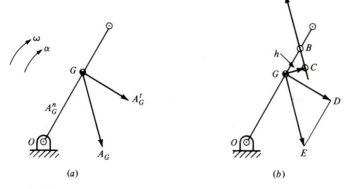

(a)   (b)

FIGURE 5.12
Center of percussion.

The distance GB is seen to be totally independent of the angular velocity or angular acceleration of the link. The point B is known as the center of percussion of the link. The inertia force will always pass through the center of percussion of a link which is rotating about a fixed center. If the link is not rotating about a fixed center (like a coupler link), the inertia force may or may not pass through the center of percussion.

The mass moment of inertia of a body is defined as $I = \Sigma\, r^2\, dM$ or as $I = Mk^2$ in which k is known as the radius of gyration. Considering then Eq. (5.1),

$$(GB)(OG) = \frac{I}{M} = k^2$$

This provides a convenient means of determination of the mass moment of inertia of a body. If a body is suspended and allowed to oscillate as a simple pendulum as shown in Fig. 5.13, the period of vibration will be

$$\tau = 2\pi\sqrt{\frac{1}{g}}$$

where τ = time for one complete oscillation

  l = distance from point of support to center of percussion, $OG + GB$

  g = acceleration of gravity

Then
$$l = OG + GB = \frac{\tau^2 g}{4\pi^2}$$

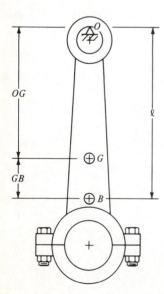

FIGURE 5.13
Determination of mass moment of inertia.

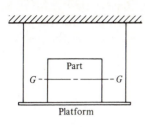
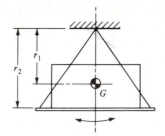

$$I_p = M_p gr[(\tau_1/2\pi)^2 - r_1/g] + Mgr_2/4\pi^2(\tau_1^2 - \tau_2^2)$$

where I_p = mass moment of inertia of part
M_p = mass of part
M = mass of platform
r_1 = distance from support point to center of gravity of part
r_2 = distance from support point to center of gravity of platform
τ_1 = period of vibration of platform alone
τ_2 = period of vibration of platform and part

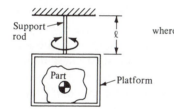

$$I_p = k/4\pi^2(\tau_2^2 - \tau_1^2)$$

where I_p = moment of inertia of part
τ_1 = period of platform alone
τ_2 = period of platform and part
k = torsional spring constant of support rod = JG/ℓ

FIGURE 5.14
Mass moment of inertia of irregular shapes. (See E. E. Crede "Determining Moment of Inertia," *Machine Design*, August 1948.)

since
$$k^2 = (GB)(OG) = (OG)(l - OG)$$

and
$$I = Mk^2 = M\left[(OG)\left(\frac{\tau^2 g}{4\pi^2} - OG\right)\right] \qquad (5.2)$$

The shape of the machine parts sometimes makes it very difficult to calculate the mass moment of inertia. With the use of Eq. (5.2) and a model of the part, the mass moment of inertia may be found experimentally. Location of the center of gravity (distance OG) may be found by several experimental means.

In addition to suspending the machine part as a simple pendulum, it is possible to construct a platform suspended to operate as a torsional or simple pendulum and measure its period of vibration. Then the part whose mass moment of inertia is to be determined is placed on the platform (with its center of gravity directly above the center of gravity of the platform) and the period of vibration is again measured. The two observations, one without the part in question and one with the part included, are used to determine the mass moment of inertia of the part as shown in Fig. 5.14.

5.6 GRAPHICAL INERTIA FORCE ANALYSIS

With the adoption of d'Alembert's principle in which the inertia forces are considered as those resulting from acceleration, each link's inertia force is placed in a direction opposite to the acceleration of the center of gravity of the link and offset from the center of gravity in such a manner as to result in a torque about the center of gravity in a sense opposite to the sense of the angular acceleration. With adoption of d'Alembert's principle, each link may be considered to be in a state of static equilibrium.

The mechanism shown in Fig. 5.15a is to be analyzed for all inertia forces and torques. The mechanism has a CCW angular velocity of the crank (link 1) of 4 rad/s which is decreasing with an angular acceleration of 5 rad/s². All measurements of the mechanism are in the conventional English system. If the mechanism measurements were to be in the metric system, the technique would not change, but care must be taken to assure dimensional stability in the units of mass and mass moment of inertia. In this example, weights of the individual links are given in pounds and the mass moments of inertia are in lb · in · s² units.

The velocity and acceleration polygons are shown in Fig. 5.15b. It will be necessary to know the angular acceleration of links 2 and 3. Therefore,

$$\alpha_2 = \frac{A^t_{B/A}}{AB} = \frac{32.0}{2.75} = 11.64 \text{ rad/s}^2 \text{ CCW}$$

$$\alpha_3 = \frac{A^t_B}{O_4B} = \frac{13.0}{2.0} = 6.50 \text{ rad/s}^2 \text{ CCW}$$

With the acceleration of the center of gravity and the weight (W) of the link known, the inertial force associated with the link may be calculated.

$$F_1 = \frac{W_1}{g} A_{G1} = \frac{3.0}{386} 12.5 = 0.097 \text{ lb}$$

$$F_2 = \frac{W_2}{g} A_{G2} = \frac{5.0}{386} 12.5 = 0.162 \text{ lb}$$

$$F_3 = \frac{W_3}{g} A_{G3} = \frac{4.0}{386} 18.0 = 0.187 \text{ lb}$$

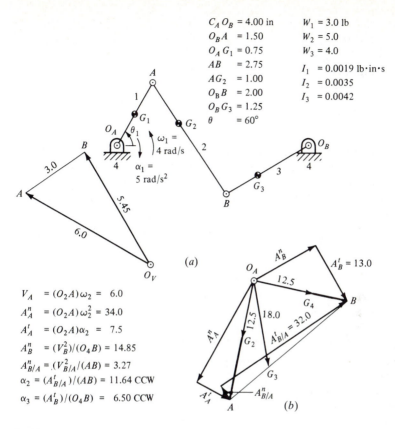

$$C_A O_B = 4.00 \text{ in} \qquad W_1 = 3.0 \text{ lb}$$
$$O_B A = 1.50 \qquad W_2 = 5.0$$
$$O_A G_1 = 0.75 \qquad W_3 = 4.0$$
$$AB = 2.75$$
$$AG_2 = 1.00 \qquad I_1 = 0.0019 \text{ lb·in·s}$$
$$O_B B = 2.00 \qquad I_2 = 0.0035$$
$$O_B G_3 = 1.25 \qquad I_3 = 0.0042$$
$$\theta = 60°$$

$$\omega_1 = 4 \text{ rad/s}$$
$$\alpha_1 = 5 \text{ rad/s}^2$$

$$(a)$$

$$V_A = (O_2 A)\omega_2 = 6.0$$
$$A_A^n = (O_2 A)\omega_2^2 = 34.0$$
$$A_A^t = (O_2 A)\alpha_2 = 7.5$$
$$A_B^n = (V_B^2)/(O_4 B) = 14.85$$
$$A_{B/A}^n = (V_{B/A}^2/(AB) = 3.27$$
$$\alpha_2 = (A_{B/A}^t)/(AB) = 11.64 \text{ CCW}$$
$$\alpha_3 = (A_B^t)/(O_4 B) = 6.50 \text{ CCW}$$

$$A_B^t = 13.0$$

$$(b)$$

FIGURE 5.15
Graphical inertia force analysis.

With the mass moment of inertia, inertia force magnitude, and angular acceleration of the link known, the offset of the inertia force from the center of gravity may be calculated.

$$h_1 = \frac{I_1 \alpha_1}{F_1} = \frac{(0.00019)(5.0)}{0.097} = 0.097 \text{ in}$$

$$h_2 = \frac{I_2 \alpha_2}{F_2} = \frac{(0.0035)(11.64)}{0.162} = 0.251 \text{ in}$$

$$h_3 = \frac{I_3 \alpha_3}{F_3} = \frac{(0.0042)(6.50)}{0.187} = 0.146 \text{ in}$$

The inertia forces are shown on the mechanism in Fig. 5.16a.

In conducting an inertia force analysis of a mechanism, it is important to know the directions of forces acting. For this reason, the

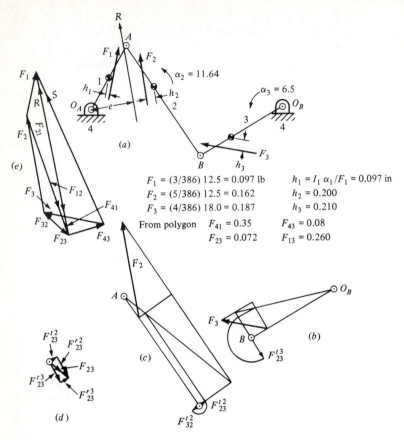

$F_1 = (3/386)\ 12.5 = 0.097\ \text{lb}$ $h_1 = I_1\,\alpha_1/F_1 = 0.097\ \text{in}$
$F_2 = (5/386)\ 12.5 = 0.162$ $h_2 = 0.200$
$F_3 = (4/386)\ 18.0 = 0.187$ $h_3 = 0.210$

From polygon $F_{41} = 0.35$ $F_{43} = 0.08$
$F_{23} = 0.072$ $F_{13} = 0.260$

FIGURE 5.16
Graphical inertia force analysis.

following nomenclature system is followed. All forces are indicated with a single subscript if that force is the result of acceleration of the link. The forces acting within the links are indicated with two subscripts such as $\mathbf{F}_{34}$ which indicates the force that link 3 exerts on link 4. The force $\mathbf{F}_{43}$ would be equal and opposite to the force $\mathbf{F}_{34}$. Recall that the use of d'Alembert's principle results in placing all links in a state of static equilibrium known as kinetostatic equilibrium.

For a complete inertia force analysis of a mechanism, it is necessary to start with a link that is far removed from the driving link since an unknown torque is acting on the driving link and summation of torques would result in excessive unknowns in an equation. Therefore, link 3 is shown as a free body in Fig. 5.16b. The forces acting on link 3 are $\mathbf{F}_3$, $\mathbf{F}_{43}$, and $\mathbf{F}_{23}$. Since the link is in a state of static equilibrium and the three forces acting are all in the same plane, the line of action of the three forces must intersect at the same point. This is known as a concurrent coplaner force system. The force $\mathbf{F}_3$ is known completely in magnitude, direction, and location. The force $\mathbf{F}_{43}$ must act through point O_B where link 3 is connected to link 4 with a revolute joint. Nothing is known about the magnitude or direction of $\mathbf{F}_{43}$. The force $\mathbf{F}_{23}$ must act through the revolute connection of links 2 and 3. Nothing is known about its magnitude or direction. However, since link 3 is in a state of static equilibrium, the sum of moments about any point must be zero. In graphically summing moments about point O_B of link 3, the force $\mathbf{F}_3$ is placed with its origin on the centerline of link 3, and it is resolved into two components, one perpendicular to link 3 and the other radial on link 3. The radial component of $\mathbf{F}_3$ has no moment about point O_B. The tangential component of $\mathbf{F}_3$ has a moment about O_B which must be opposed by the force $\mathbf{F}_{23}$ acting through point B. Whatever the magnitude and direction of $\mathbf{F}_{23}$, it must have a component perpendicular to link 3 ($\mathbf{F}_{23}^{t3}$) such that its moment about O_B will be equal to F_3^{t3} multiplied by its distance from O_B. A graphical means for finding $\mathbf{F}_{23}^{t3}$ is shown in Fig. 5.16b.

Link 2 is shown as a free body in Fig. 5.16c. The forces acting on link 2 are $\mathbf{F}_2$, $\mathbf{F}_{32}$, and $\mathbf{F}_{12}$. Since link 2 is in a state of static equilibrium, the sum of moments about point A must be zero. In summing moments, the force $\mathbf{F}_2$ is placed with its origin on the centerline of link 2 and is resolved into two components, one perpendicular to link 2 and the other radial on link 2. The radial component has no moment about point A. The moment resulting from the tangential component must be balanced by the moment resulting from the force $\mathbf{F}_{32}$. Whatever the force $\mathbf{F}_{32}$ is in magnitude and direction, it must have the component $\mathbf{F}_{32}^{t2}$ as found graphically in Fig. 5.16c. The force $\mathbf{F}_{23}$ is equal and opposite to the force $\mathbf{F}_{32}^{t2}$.

Now the magnitude and direction of the force $\mathbf{F}_{23}$ may be found by the simultaneous solution of two equations as shown in Fig. 5.16d. The equations to be solved are

$$\mathbf{F}_{23} = \mathbf{F}_{23}^{t3} + \mathbf{F}_{23}^{r3}$$

and

$$\mathbf{F}_{23} = \mathbf{F}_{23}^{t2} + \mathbf{F}_{23}^{r2}$$

The force $\mathbf{F}_{43}$ passes through the intersection of the lines of action of forces $\mathbf{F}_3$ and $\mathbf{F}_{23}$.

The forces acting on link 3 may be placed into a force polygon as in Fig. 5.16e. Starting with the force $\mathbf{F}_3$ and adding the force $\mathbf{F}_{23}$, then force $\mathbf{F}_{43}$ completes the polygon for link 3. As a check, the line of action of $\mathbf{F}_{43}$ found from the polygon should be checked against the line of action of $\mathbf{F}_{43}$ which must pass through the intersection of $\mathbf{F}_3$ and $\mathbf{F}_{23}$ and the pivot point O_B.

This completes consideration of link 3 of the mechanism, and attention is now directed to link 2. In the force polygon of link 3, the force $\mathbf{F}_{23}$ may be reversed to produce the force $\mathbf{F}_{32}$. The other forces acting on link 2 are $\mathbf{F}_2$ and $\mathbf{F}_{12}$. These three forces are in a state of static equilibrium and together must form a closed force polygon. Therefore, the force vector $\mathbf{F}_2$ may be added to the force vector $\mathbf{F}_{32}$ in the force polygon. The vector from the head of the $\mathbf{F}_2$ force vector back to the origin of the $\mathbf{F}_{32}$ force vector will be the force vector $\mathbf{F}_{12}$. The line of action of the force $\mathbf{F}_{12}$ must be through the point A to the intersection of the lines of action of the forces $\mathbf{F}_2$ and $\mathbf{F}_{32}$.

Forces acting on link 1 are $\mathbf{F}_1$, $\mathbf{F}_{21}$, $\mathbf{F}_{41}$, and a torque produced or absorbed by the driver. In the force polygon, the force vector $\mathbf{F}_{12}$ may be reversed to become the force vector $\mathbf{F}_{21}$. If the force vector $\mathbf{F}_2$ is added to the terminus of the force vector $\mathbf{F}_{21}$, the polygon will be closed by the force vector $\mathbf{F}_{41}$.

Since there may be a torque added to or resisted through the crank pivot O_A, the lines of action of the forces $\mathbf{F}_1$, $\mathbf{F}_{21}$, and $\mathbf{F}_{41}$ do not necessarily need to intersect at one point. The forces $\mathbf{F}_1$ and $\mathbf{F}_{21}$ may be added to form one force $\mathbf{R}$ which acts through the intersection of the lines of action of the forces $\mathbf{F}_1$ and $\mathbf{F}_{21}$. The force $\mathbf{R}$ will be equal and opposite to the force $\mathbf{F}_{41}$. If an external torque is acting on the link, the line of action of the force $\mathbf{R}$ will not pass through the point O_A but will be offset an amount e. The torque acting will be in a sense opposite to the torque produced by force $\mathbf{R}$ at a distance e and will be equal to force $\mathbf{R}$ multiplied by perpendicular distance e.

Note that in the completed force polygon the sum of forces $\mathbf{F}_3$, $\mathbf{F}_2$, $\mathbf{F}_1$, $\mathbf{F}_{41}$, and $\mathbf{F}_{43}$ will be equal to zero. Also the sum of the inertia forces $\mathbf{F}_3$, $\mathbf{F}_2$, and $\mathbf{F}_1$ will be equal to $\mathbf{S}$ (the total shaking force for the mechanism).

If the mechanism is considered to be contained in a box, the force $\mathbf{S}$ will attempt to move the box while the forces $\mathbf{F}_{41}$ and $\mathbf{F}_{43}$ will be the hold-down forces which resist motion of the box.

5.7   ANALYTIC INERTIA FORCE ANALYSIS

Calculation of the inertia forces acting on a mechanism is complicated by the requirement of determining the point of application of a single inertia force on each link. This requires that the force magnitude, its point of application, and its direction be calculated.

If the inertia forces acting on a link are considered as a single force acting through the center of gravity together with an inertia torque, the inertia force analysis of a mechanism is greatly simplified. The procedure requires solution of simultaneous equations, and some knowledge of matrix manipulation as given in App. B may be helpful. The developed simultaneous equations may be solved using a recognized technique for solution of simultaneous equations which is provided in program SIMULT in App. A.

The mechanism of Fig. 5.17 is used as an example of the techniques. Known quantities include:

1. Configuration of the mechanism

$$r_1, r_2, r_3, r_4, \theta_1, \theta_2, \theta_3$$

2. Location of the center of gravity of each link

$$RG_1, R_1, R_2, RG_3$$

$$\epsilon = \arctan \frac{R_2}{R_1}$$

3. Acceleration of the center of gravity of each link

$$\mathbf{A}_{G1}, \mathbf{A}_{G2}, \mathbf{A}_{G3} \text{ in magnitude and direction}$$

4. Angular acceleration of each link

$$\alpha_1, \alpha_2, \alpha_3 \text{ in magnitude and direction}$$

5. Mass of each link

$$M_1, M_2, M_3$$

6. Mass moment of inertia of each link

$$I_1, I_2, I_3$$

With the consideration of kinetostatic equilibrium, each link may be considered as a separate free body, and the sum of forces in any direction as well as the sum of moments about any point will be equal to zero. It is important to remember that the forces being considered are due to acceleration of the members.

Figure 5.17a shows the driving link as a free body with the inertia force $\mathbf{F}_1$ resolved into x and y components. The forces $\mathbf{F}_{41}^x$ and $\mathbf{F}_{41}^y$ are the x and y components of the force that link 4 exerts on link 1. In like manner the forces $\mathbf{F}_{21}^x$ and $\mathbf{F}_{21}^y$ are the x and y components of the force that link 2 exerts on link 1. The torque $\mathbf{T}_1$ is the inertia torque associated with the angular acceleration of link 1. The torque $\mathbf{T}_i$ is the input torque required by the driver of the system in order to maintain the indicated accelerations.

Summation of forces in the x and y directions as well as summation of moments (torques) about the center of gravity of the link provides three equations:

$$\mathbf{F}_1^x = \mathbf{F}_{41}^x + \mathbf{F}_{21}^x \tag{5.3}$$

$$\mathbf{F}_1^y = \mathbf{F}_{41}^y + \mathbf{F}_{21}^y \tag{5.4}$$

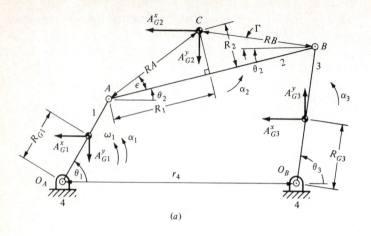

(a)

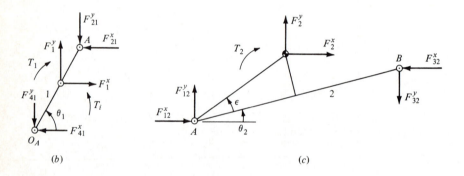

(b)

(c)

(d)

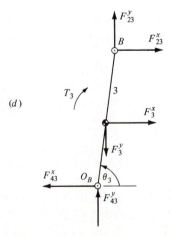

FIGURE 5.17
Inertia forces.

$$T_1 = -T_i + F_{41}^x R_{G1} \sin \theta_1 - F_{41}^y R_{G1} \cos \theta_1 + F_{21}^x (r_1 - R_{G1}) \sin \theta_1$$
$$-F_{21}^y (r_2 - R_{G1}) \cos \theta_1 \tag{5.5}$$

In sketching link 2 as a free body (Fig. 5.17c) it is important to recognize that $\mathbf{F}_{ij} = -\mathbf{F}_{ji}$. Care must be taken to ensure that the direction of the forces is consistent from one free-body diagram to another. After the free-body diagram is sketched, summation of forces and moments will produce three equations for link 2.

$$\mathbf{F}_2^x = -\mathbf{F}_{12}^x + \mathbf{F}_{32}^x \tag{5.6}$$

$$\mathbf{F}_2^y = -\mathbf{F}_{12}^y + \mathbf{F}_{32}^y \tag{5.7}$$

$$T_2 = -F_{12}^x RA \sin(\theta_2 + \epsilon) + F_{12}^y RA \cos(\theta_2 + \epsilon) + F_{32}^x RB \sin \Gamma$$
$$-F_{32}^y RB \cos \Gamma \tag{5.8}$$

Link 3 is shown as a free body in Fig. 5.17d. Again summation of forces and moments will produce three equations for link 3.

$$\mathbf{F}_3^x = \mathbf{F}_{43}^x - \mathbf{F}_{23}^x \tag{5.9}$$

$$\mathbf{F}_3^y = \mathbf{F}_{43}^y + \mathbf{F}_{23}^y \tag{5.10}$$

$$T_3 = -F_{43}^x R_{G3} \sin \theta_3 - F_{43}^y R_{G3} \cos \theta_3 - F_{23}^x (r_3 - R_{G3}) \sin \theta_3$$
$$+F_{23}^y (r_3 - R_{G3}) \cos \theta_3 \tag{5.11}$$

The left sides of Eqs. (5.3) through (5.11) are all known quantities. On the right side are the unknown x and y components of the pin forces and the input torques. Solution of Eqs. (5.3) through (5.11) provides the x and y components of all pin forces and hold-down forces in the mechanism.

Example 5.1. Known parameters:

1. Configuration of the mechanism

$$r_4 = 4.0 \qquad r_1 = 1.5 \qquad r_2 = 2.75 \qquad r_3 = 2.0$$
$$\theta_1 = 60° \qquad \theta_2 = 303.5° \qquad \theta_3 = 211°$$

2. Location of the center of gravity (Fig. 5.14)

$$RG_1 = 0.75 \qquad R_1 = 1.0 \qquad R_2 = 0 \qquad r_2 = 1.75$$
$$RG_3 = 1.25$$

3. Acceleration of centers of gravity

$$A_{G1} = 1.25 \text{ at } 240.5°$$
$$A_{G2} = 15.5 \text{ at } 280°$$
$$A_{G3} = 12.5 \text{ at } 348.5°$$

4. Angular acceleration of each link

$$\alpha_1 = 5.00 \text{ CW}$$

$$\alpha_2 = 11.64 \text{ CCW}$$

$$\alpha_3 = 6.50 \text{ CCW}$$

5. Mass and moment of inertia of each link

$$M_1 = 0.00777 \text{ lb} \cdot \text{s}^2/\text{in} \qquad I_1 = 0.0019 \text{ lb} \cdot \text{in} \cdot \text{s}^2$$

$$M_2 = 0.01295 \text{ lb} \cdot \text{s}^2/\text{in} \qquad I_2 = 0.0035 \text{ lb} \cdot \text{in} \cdot \text{s}^2$$

$$M_3 = 0.01036 \text{ lb} \cdot \text{s}^2/\text{in} \qquad I_3 = 0.0042 \text{ lb} \cdot \text{in} \cdot \text{s}^2$$

With the known quantities, the free-body diagrams of each link will be as shown in Fig. 5.17, and the equations of kinetostatic equilibrium (with care taken to match the direction of the forces with those used in derivation of the equations) become:

$$0.0478 = -\mathbf{F}_{41}^x + \mathbf{F}_{21}^x$$

$$0.0844 = \mathbf{F}_{41}^y + \mathbf{F}_{21}^y$$

$$0.0095 = \mathbf{T}_i + 0.6528\mathbf{F}_{41}^x + 0.3693\mathbf{F}_{41}^y$$

$$+ 0.6528\mathbf{F}_{21}^x - 0.3693\mathbf{F}_{21}^y$$

$$-0.0349 = -\mathbf{F}_{12}^x - \mathbf{F}_{32}^x$$

$$0.1977 = -\mathbf{F}_{12}^y - \mathbf{F}_{32}^y$$

$$-0.0407 = +0.8339\mathbf{F}_{12}^x + 0.5519\mathbf{F}_{12}^y$$

$$+ 0.8339\mathbf{F}_{32}^x + 0.5519\mathbf{F}_{32}^y$$

$$-0.1269 = \mathbf{F}_{43}^x - \mathbf{F}_{23}^x$$

$$0.0258 = \mathbf{F}_{43}^y - \mathbf{F}_{23}^y$$

$$-0.0273 = -0.6438\mathbf{F}_{43}^x - 1.0716\mathbf{F}_{43}^y$$

$$-0.3863\mathbf{F}_{23}^x - 0.6429\mathbf{F}_{23}^y$$

These nine equations with nine unknowns may be solved using the program SIMULT (App. A). It should be noted that reversing the subscripts on a force also reverses the sign of the term in the equation. In the event that some of the forces result in negative numbers, that force is pointed in the opposite direction selected in deriving the equations.

5.8 COMPUTER PROGRAM FOR ACCELERATION OF CENTERS OF GRAVITY

The acceleration program may be modified to provide the acceleration of the centers of gravity of the input crank and output rocker as given by CGACCELERATION (App. A).

With the acceleration of the centers of gravity known, the inertia force analysis can proceed either graphically or analytically.

5.9 BALANCING

A machine component, or link, may be balanced either statically or dynamically. Static balance is accomplished by adding or removing weights (or relocating weights) until the component will remain static when placed on knife edges as shown in Fig. 5.18. Added weights are known as counterweights. Static balance of machine components is not sufficient to remove inertia forces since the forces may act in various planes or in various directions depending on the direction and magnitude of the acceleration.

It is not uncommon for the inertia forces of mechanisms to result in very large shaking forces. This is particularly true for lightweight, high-speed machinery or for slow-moving, very heavy machinery. Inertia forces can result in excessive deflection of the links, very high stress in the components producing failure of the links, or improper coupler curve generation.

The designer must balance the mechanism dynamically. Balancing is accomplished either by reducing the mass or acceleration of links or by introducing forces opposite in direction of the inertia forces. Since it is often very difficult to reduce mass or acceleration of the links, the addition of forces to counteract the inertia forces is the most attractive method for reducing shaking forces.

Figure. 5.19 shows a rotating shaft which is balanced statically but which would exhibit very large shaking forces if it is rotated at high speed. The unbalanced forces which cause a shaking moment exist since the inertia forces act in different planes. The magnitude of the shaking moment will increase with the square of the angular velocity and directly with the angular acceleration of the shaft. Dynamic balance can be achieved by adding mass to the system such that the inertia forces resulting from the added mass will be equal and opposite to those causing the shaking moment.

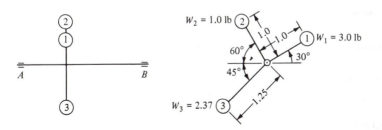

$$W_1(1.0 \cos 30) = W_2(1.0 \cos 60) + W_3(1.25 \cos 45)$$

FIGURE 5.18
Static balance.

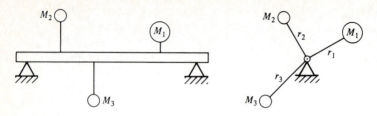

FIGURE 5.19
Static but not dynamic balance.

In most cases it is not possible to add mass to the rotating system in the plane of the shaking forces. In this case, a shaking moment may be added to counteract the disturbing moment. The added shaking moment may be found either graphically or analytically.

Figure 5.20a shows a shaft with three unbalanced masses which will produce inertia forces acting in the planes of the unbalanced masses. When operating at constant speed, the inertia forces associated with each mass will be equal to $Mr\omega^2$ and will be directed radial outward in the

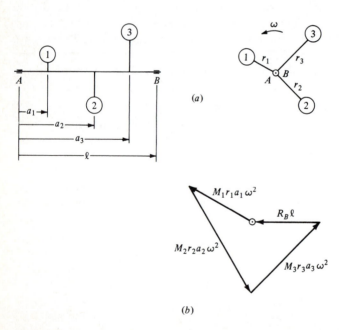

(a)

(b)

FIGURE 5.20
Dynamic bearing force.

direction of the unbalanced mass. The unbalanced shaking forces will produce additional forces on the shaft bearings. By summing moments about one bearing the magnitude and direction of force at the other bearing may be determined. Realize that the direction of the bearing force is rotating at the same angular velocity as the shaft.

Summing moments about the left bearing (Fig. 5.20) will provide the right bearing force:

$$R_B(l) + M_3 r_3 \omega^2 a_3 + M_2 r_2 \omega^2 a_2 + M_1 r_1 \omega^2 a_1 = 0$$

A graphical solution of this equation is shown in Fig. 5.20b. An analytical solution will require either adding components of the forces in the x and y directions or treating the inertia forces as vectors and solving the equation vectorially. A similar equation written as moments about the right bearing will provide the left bearing force. The designer is now able to decide whether the bearing forces are acceptable or not. If the bearing forces and resulting induced shaft stress are excessive, the designer must dynamically balance the system.

An unbalanced shaft is shown in Fig. 5.21a and is to be dynamically balanced by adding mass in planes A and B of the shaft. Location of planes A and B is determined by space limitations of the system being designed. The amount of mass and its orientation in each of the planes is to be determined. If the system is completely balanced dynamically, the dynamic bearing forces will be zero. Summing moments about plane A of Fig. 5.21 provides information for the mass and its location in plane B.

$$M_1 r_1 \omega^2 a_1 + M_2 r_2 \omega^2 (-a_2) + M_3 r_3 \omega^2 (-a_3) + M_B r_B \omega^2 a_B = 0$$

Note that ω^2 is common to all terms of the equation and can be eliminated. This means that the system will be dynamically balanced at all angular velocities. Note also that the direction of the shaking moment is taken into account by the sign of the distance. Since the system is to be dynamically balanced, the bearing forces are not considered.

The angular accelerations may produce some shaking forces but at steady-state operation the system will be dynamically balanced. The equation is solved graphically in Fig. 5.21b resulting in the value of $M_B r_B$. The mass M_B may be large and located at a small radius from the center of rotation, or the mass M_B may be small and located at a large radius. The designer may adjust the amount of mass and its location depending upon the machine configuration. However, the product of the mass M_B and its radius r_B has been determined. Summing moments about plane B will provide $M_A r_A$.

Analytic determination of the balancing mass and its radius from the center of rotation is generally done by resolving the shaking forces into x and y components.

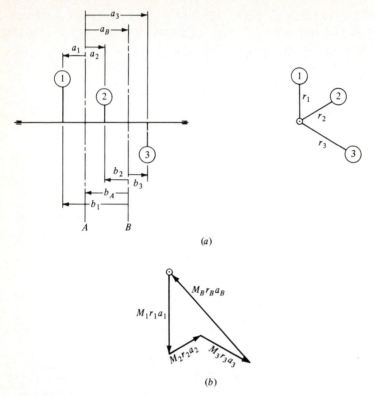

(a)

(b)

FIGURE 5.21
Balancing mass determination.

5.10   BALANCING THE SLIDER CRANK MECHANISM

Balancing of the slider crank and similar mechanisms presents problems since the unbalanced masses are not simply rotating but are also reciprocating and/or moving in various coupler curve paths. It is generally not possible to completely balance such mechanisms, and the designer is faced with the problem of reducing the unbalanced shaking forces to an acceptable minimum.

Shaking forces associated with the slider crank mechanism shown in Fig. 5.22a may be computed analytically by treating each link separately. In conducting the analysis it is convenient to relocate the accelerating masses considered located at the center of gravity to more easily handled, yet dynamically equivalent, locations.

Considering the crank alone (Fig. 5.22) with constant angular velocity the inertia force will be equal to $M_1 r \omega^2$ directed radially outward

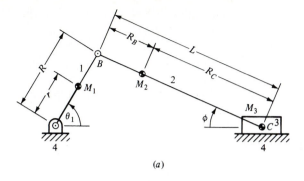

(a)

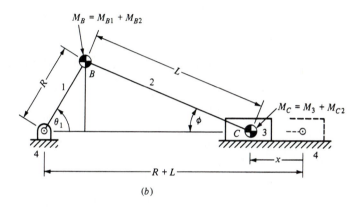

(b)

FIGURE 5.22
Dynamically equivalent slider crank.

from the center of rotation. An equivalent inertia force will result from a mass equal to M_{B1} located at radius R or at point B.

$$M_{B1} = M_1 \frac{r}{R}$$

M_{B1} is known as a dynamically equivalent mass of the crank and is located at the crank pin. This is a dynamically equivalent mass only for a crank which is rotating at constant angular velocity. Note that the equivalent mass M_{B1} is slightly less than the true mass M_1 of the crank. Since the equivalent mass is located at a larger radius from the center of rotation, the magnitude and direction of the inertia force resulting from use of the equivalent mass are correct. Thus,

$$\mathbf{F}_1 = M_1 r \boldsymbol{\omega}^2 = M_{B1} R \boldsymbol{\omega}^2 \qquad \text{directed radially outward from the center of rotation}$$

Considering the coupler link (connecting rod) alone (Fig. 5.22), the direction of the acceleration and thus the direction of the associated inertia force is constantly changing. It would be most convenient if the mass of the connecting rod could be replaced by one or more masses located where the direction of acceleration is more easily determined. In replacing the original mass with equivalent masses several important factors must be considered:

1. The total mass of the link must remain unchanged.
2. The location of the center of gravity must remain unchanged.
3. The mass moment of inertia of the link must remain unchanged.

If the total mass of the coupler is to be replaced by two masses, the most convenient locations of the equivalent masses are one at the crank pin (point B) and the other at the piston (point C).

The total mass of the link will be unchanged if the sum of the two equivalent masses is equal to the mass of the link.

$$M_2 = M_{B2} + M_{C2} \qquad (a)$$

The center of gravity location will be unchanged if the two equivalent masses lie on a straight line through the center of gravity and if the sum of the moments of the equivalent masses about the center of gravity of the link is zero.

$$M_{B2}R_B - M_{C2}R_C = 0 \qquad (b)$$

The moment of inertia of the link will be unchanged if the sum of the moments of inertia of the equivalent masses about the center of gravity is equal to the original moment of inertia of the link.

$$M_{B2}R_B + M_{C2}R_C = I \qquad (c)$$

Simultaneous solution of Eqs. (a) and (b) provides

$$M_{C2} = M_2 \frac{R_B}{R_C + R_B} \qquad (d)$$

$$M_{B2} = M_2 \frac{R_C}{R_C + R_B} \qquad (e)$$

Substitution of Eqs. (d) and (e) into Eq. (c) results in

$$M_2 R_B R_C = I \qquad (f)$$

Equations (d), (e), and (f) are three equations with four unknowns: M_{B2}, M_{C2}, R_C, and R_B. Selection of one of the unknowns allows for solution of the other three unknowns. Unfortunately, the distances R_B and R_C are not really unknowns. If R_C is selected as a known quantity, then R_B will become some value different from that desired by a small amount and the equivalent mass M_2 is not located at the crank pin. This difference is generally neglected and will introduce a small error in dynamic analysis of the slider crank system. If the error is intolerable, it is necessary to follow the procedures of Sec. 5.6 or 5.7.

By using the equivalent mass technique, the slider crank mechanism is reduced to one of two masses (Fig. 5.22). One mass is considered to be concentrated at the crank pin and consists of the equivalent mass of the crank plus a part of the equivalent mass of the connecting rod. The other mass is considered to be concentrated at the piston and consists of the piston mass plus a part of the equivalent mass of the connecting rod.

Dynamic analysis of the equivalent mass slider crank mechanism may be conducted analytically and/or graphically. The total shaking force will be due to the two equivalent masses. It is necessary to compute the shaking forces and to add them vectorially.

The shaking force due to the mass at the crank pin is simply

$$\mathbf{F}_C = M_B R \omega_1^2 \qquad \text{directed radially outward along the crank}$$

The shaking force due to the mass at the piston is equal to the mass at the piston multiplied by its acceleration. With reference to Fig. 5.22, displacement of the piston will be

$$X = (R + L) - R \cos \theta_1 - L \cos \phi$$

$$= R(1 - \cos \theta_1) + L(1 - \cos \phi)$$

Note:
$$R \sin \theta_1 = L \sin \phi$$

$$\cos^2 \phi = 1 - \frac{R^2}{L^2} \sin^2 \theta_1$$

Then

$$X = R(1 - \cos \theta_1) + L\left(1 - \sqrt{1 - \frac{R^2}{L^2} \sin^2 \theta_1}\right)$$

In order to simplify differentiation of the displacement equation, the last term is expanded and only the first two terms of a binomial series are used. (Greater accuracy results by using more terms of the expansion.) Then

$$\sqrt{1 - \frac{R^2}{L^2} \sin^2 \theta_1} = 1 - \frac{1}{2}\left(\frac{R}{L}\right)^2 \sin^2 \theta_1$$

Because the angular velocity of the crank is constant and $d\theta/dt = \omega$, differentiations of the displacement equation give the velocity and the acceleration of the piston.

$$\mathbf{V}_x = \frac{dx}{dt} = R\omega_1\left(\sin \theta_1 + \frac{R}{2L} \sin 2\theta_1\right)$$

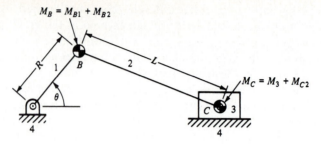

$$S = M_B R\omega_1^2 + M_C R\omega_1^2 \cos\theta + M_C R\omega_1^2 (R/L) \cos 2\theta$$

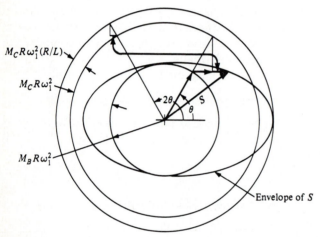

FIGURE 5.23
Polar shaking force diagram.

$$\mathbf{A}_X = \frac{d^2x}{dt^2} = R\omega_1^2\left(\cos\theta_1 + \frac{R}{L}\cos 2\theta_1\right)$$

The inertia force associated with the piston becomes

$$\mathbf{F}_P = M_C R\omega_1^2\left(\cos\theta_1 + \frac{R}{L}\cos 2\theta_1\right)$$

The total shaking force associated with the slider crank mechanism is the vector sum of the crank and piston shaking forces.

$$\mathbf{S} = M_B R\omega_1^2 + M_C R\omega_1^2 \cos\theta_1 + M_C R\omega_1^2 \frac{R}{L}\cos 2\theta_1$$

In this equation, the first term is the crank shaking force, the second term is the primary shaking force, and the third term is the secondary shaking force. The equation may be solved graphically in a polar shaking force diagram as shown in Fig. 5.23.

Addition of a counterweight to the system will add another shaking force and will have a pronounced influence on the total shaking force of

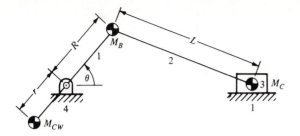

$$S = M_B R\omega_1^2 + M_C R\omega_1^2 \cos\theta + M_C R\omega_1^2 (R/L) \cos 2\theta + M_{CW} r\omega_1^2$$

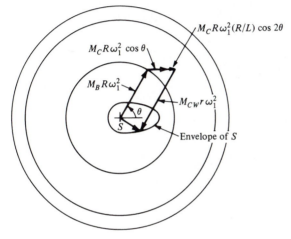

FIGURE 5.24
Polar shaking force diagram with counterweight.

the system. The counterweight of Fig. 5.24 is arranged to rotate with the crank, and its inertia force is directed opposite to the crank shaking force.

PROBLEMS

5.1. Determine the force of the piston (link 1) necessary to maintain static equilibrium with **F** = 2000.

$$O_B B = 3.4 \qquad O_A B = 3.75 \qquad O_A A = 8.0$$

FIGURE P5.1
Static force analysis.

5.2. The mechanism shown is in a state of static equilibrium. Determine the forces acting in all the revolute joints and the torque to be applied to link 1.

$$O_A O_B = 3.0 \qquad O_A A = 1.0 \qquad O_B B = 1.5 \qquad F = 2000$$
$$O_B C = CD = 0.75$$

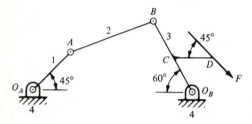

FIGURE P5.2
Static force analysis.

5.3. The wall-mounted crane is designed to lift 2.5 tons. Determine the required torque to be applied to link 3.

$$O_A A = O_B B = 6.5'$$
$$AC \;\; = 8.0$$

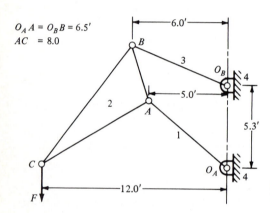

FIGURE P5.3
Wall-mounted crane.

5.4. A force **F** of 1500 lb acts on the piston of the slider crank mechanism causing impending motion of the piston to the left.
(*a*) Determine the torque applied to link 1.
(*b*) With coefficient of friction equal to 0.17 and all pin diameters equal to 1.5, determine the torque applied to link 1.

$$O_A A = 2.0'' \qquad AB = 4.0''$$

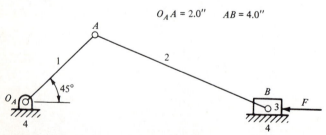

FIGURE P5.4
Slider crank force analysis with and without friction.

5.5. Using the pole force technique, determine the force **f** necessary to counterbalance the force **F**.

$$O_A O_B = 3.5 \qquad O_A A = 1.5 \qquad O_B B = AB = 2.0$$
$$O_A C = 0.75 \qquad O_B D = 1.0$$

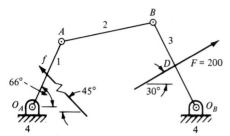

FIGURE P5.5
Pole force analysis.

5.6. Determine the piston force needed to maintain static equilibrium.

$$O_A O_B = 2.50 \qquad O_A A = 0.5 \qquad O_B B = 2.00$$
$$AB = 3.15 \qquad BC = 1.0 \qquad O_B B = 2.22$$

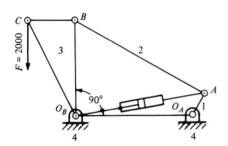

FIGURE P5.6
Pole force analysis.

5.7. Determine the force **f** needed to maintain static equilibrium with forces **F**
and **G**.

$$O_A O_B = 3.0 \qquad O_A A = 1.5 \qquad O_B B = 2.5$$
$$O_A E = 0.75 \qquad O_B D = 1.25 \qquad AC = 2.5$$

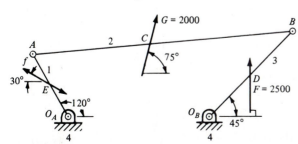

FIGURE P5.7
Pole force analysis.

5.8. The link AB has a weight of 9 lb and a mass moment of inertia about its center of gravity of $0.019 \, \text{lb} \cdot \text{ft} \cdot \text{s}^2$.

(a) Determine the radius of gyration of the link.

(b) Compute the inertia force acting on the link and show it in its proper position.

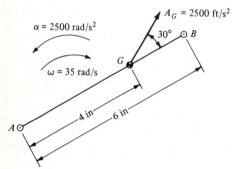

FIGURE P5.8
Radius of gyration and inertia force.

5.9. The link shown weighs 4.5 lb and has a mass moment of inertia about its center of gravity of $0.001 \, \text{lb} \cdot \text{ft} \cdot \text{s}^2$. Angular velocity of the link is 30 rad/s CCW.

(a) If the angular acceleration is zero, show the inertia force acting on the link.

(b) If the angular acceleration is 900 rad/s² CCW, show the inertia force acting on the link.

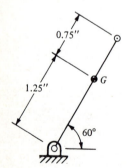

FIGURE P5.9
Crank inertia force.

5.10. The link AB has a mass of 4 kg and a mass moment of inertia about its center of gravity of $0.026 \, \text{kg} \cdot \text{m}^2$.

(a) Determine the radius of gyration of the link.

(b) Compute the inertia force acting on the link and show it in its proper position.

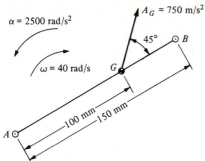

FIGURE P5.10
Floating link inertia force.

5.11. The link shown has a mass of 2 kg and a mass moment of inertia about its center of gravity of $0.0 \text{ kg} \cdot \text{m}^2$. Angular velocity of the link is 30 rad/s CCW.

(a) If the angular acceleration is zero, show the inertia force acting on the link.

(b) If the angular acceleration is 900 rad/s^2 CCW, show the inertia force acting on the link.

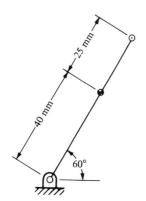

FIGURE P5.11
Crank inertia force.

5.12. Show the inertia force associated with each link in its proper position.

$O_A O_B = 10 \text{ in}$	$O_A A = 4.0 \text{ in}$	$O_A G_1 = 3.0 \text{ in}$
$AG_2 = 3.0 \text{ in}$	$O_B B = 6.0 \text{ in}$	$O_B G_3 = 3.0 \text{ in}$
$AB = 7.0 \text{ in}$		
$W_1 = 3.0 \text{ lb}$	$W_2 = 5.0 \text{ lb}$	$W_3 = 4.0 \text{ lb}$
$I_1 = 0.0014 \text{ lb} \cdot \text{ft} \cdot \text{s}^2$	$I_2 = 0.0130 \text{ lb} \cdot \text{ft} \cdot \text{s}^2$	$I_3 = 0.0046 \text{ lb} \cdot \text{ft} \cdot \text{s}^2$

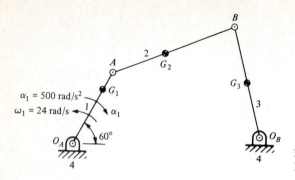

FIGURE P5.12
Inertia force analysis.

5.13. Show the inertia force associated with each link in its proper position.

$O_A O_B = 100$ mm	$O_A A = 40$ mm	$O_A G_1 = 30$ mm
$AG_2 = 30$ mm	$O_B B = 60$ mm	$O_B G_3 = 30$ mm
$AB = 70$ mm		
$M_1 = 1.50$ kg	$M_2 = 2.25$ kg	$M_3 = 2.00$ kg
$I_1 = 0.00033$ kg·m²	$I_2 = 0.00135$ kg·m²	$I_3 = 0.00103$ kg·m²

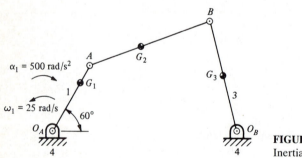

FIGURE P5.13
Inertia force analysis.

5.14. Show the inertia force for each link of the slider crank mechanism in its proper position.

$O_A A = 3.0$ in	$O_A G_1 = 2.0$ in	$AB = 8.0$ in	$AG_3 = 3.0$ in
$W_1 = 5.0$ lb	$W_2 = 8.0$ lb	$W_3 = 6.0$ lb	
$I_2 = 0.0164$ lb·ft·s²			

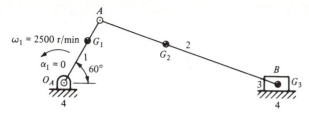

FIGURE P5.14
Slider crank inertia force analysis.

5.15. Show the inertia force for each link of the slider crank mechanism in its proper position.

$$O_A A = 40.0 \text{ mm} \qquad O_A G_1 = 20.0 \text{ mm} \qquad AB = 110.0 \text{ mm}$$

$$AG_2 = 40 \text{ mm} \qquad I_2 = 0.00065 \text{ kg} \cdot \text{m}^2$$

$$M_1 = 0.15 \text{ kg} \qquad M_2 = 0.25 \text{ kg} \qquad M_3 = 0.20 \text{ kg}$$

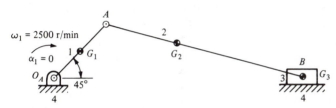

FIGURE P5.15
Slider crank inertia force analysis.

5.16. Show the inertia force associated with each link in its proper position.

$$O_C O_B = 1.75 \text{ in} \qquad O_C B = 2.0 \qquad O_C C = 3.0 \qquad O_B B = 1.0$$

$$CD = 2.0 \qquad O_B G_1 = 0.5 \qquad O_C G_3 = 1.5 \qquad CG_5 = 1.0$$

$$W_1 = 2.0 \text{ lb} \qquad W_2 = 1.0 \text{ lb} \qquad W_3 = 4.0 \text{ lb}$$

$$W_5 = 3.0 \text{ lb} \qquad W_6 = 5.0 \text{ lb}$$

$$I_3 = 0.0006 \text{ lb} \cdot \text{ft} \cdot \text{s}^2 \qquad\qquad I_5 = 0.0020 \text{ lb} \cdot \text{ft} \cdot \text{s}^2$$

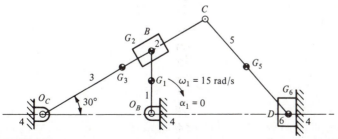

FIGURE P5.16
Inertia force analysis.

5.17. The inertia forces acting on a slider crank mechanism are given. Find the torque applied to link 1 in order to maintain equilibrium. Find the total shaking force.

$$O_AA = 4.0 \text{ in} \qquad O_AD = 2.5 \text{ in} \qquad AB = 12.0 \text{ in} \qquad AC = 5.0 \text{ in}$$

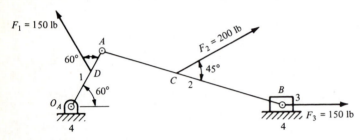

FIGURE P5.17
Inertial torque and shaking force.

5.18. With the inertia forces given determine the total shaking force and the torque required at link 1 to maintain equilibrium.

$$O_AO_B = 16.0 \text{ in} \qquad O_AA = 4.0 \text{ in} \qquad O_BB = 8.0 \text{ in} \qquad O_AC = 2.5 \text{ in}$$
$$O_BE = 5.0 \text{ in} \qquad AD = 4.0 \text{ in} \qquad AB = 10.0 \text{ in}$$

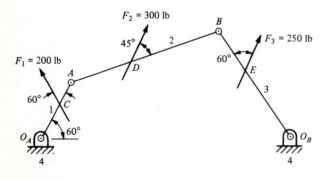

FIGURE P5.18
Inertial torque and shaking force.

5.19. For the fly ball governor which is rotating at 350 r/min, determine the total spring force at link 3 to maintain equilibrium.

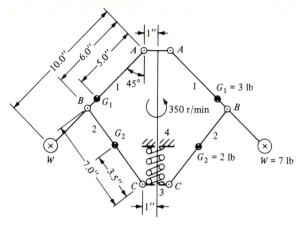

FIGURE P5.19
Fly ball governor.

5.20. Find the x and y components of all the forces of the example in Sec. 5.7.

5.21. With the unbalanced shaft rotating at 1750 r/min determine the bearing reactions at A and B.

$$W_1 = 3 \text{ lb} \qquad W_2 = 7 \text{ lb} \qquad W_3 = 5 \text{ lb}$$

Find the masses and their orientation in planes C and D in order to reduce the bearing reactions to zero.

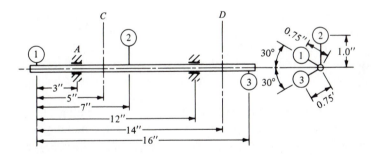

FIGURE P5.21
Bearing forces.

CHAPTER
6

GEOMETRY
OF
MECHANISMS

6.1 INTRODUCTION

The design (synthesis) of mechanisms requires extensive use of geometry in order to explain motion or to derive equations for analytic work. Several of the more useful geometric concepts are included.

Synthesis of mechanisms is a procedure by which a mechanism is developed to perform a specified task. The task specification may include:

1. Generation of a coupler curve with or without timing. Timing relates the input crank position with specified positions of the coupler point.
2. Relating input crank rotation with a specified output rocker rotation (function generation).
3. Location of crank and rocker fixed and moving pivot points to guide a coupler plane through specified positions.

Synthesis of mechanisms may be one of three principal types.

1. Dimensional synthesis is concerned with determination of the length of links.
2. Type synthesis is involved with the types of mechanisms capable of completing the desired task.

3. Number synthesis is concerned with the number of links composing the mechanism.

Of the three types of synthesis, dimensional synthesis is of most importance to the machine designer. With dimensional synthesis, it is possible to determine the size of each link and the location of fixed and moving pivots. This information is vital at the outset of a machine design. Type synthesis is very effective but requires that the designer have a very wide background and much experience in the design of mechanisms. After type synthesis is completed, the designer is still faced with the task of determining the size of each component and the pivot locations. Number synthesis leads to determination of a great many mechanisms which could perform the desired task. On completion of number synthesis, the designer is still faced with type and dimensional synthesis. Type and number synthesis techniques will help to assure that the designer has not overlooked an obvious solution to the design problem. They are particularly good techniques for brainstorming new ideas. Dimensional synthesis is direct and provides necessary information for the designer. The exclusive use of dimensional synthesis may result in overlooking equally or more effective mechanisms.

Synthesis of mechanisms may be conducted by purely graphical or purely analytic techniques or by a combination of both techniques. The graphical technique is generally very fast but requires great care in graphical skills and some would criticize the accuracy of the graphical work. The graphical techniques have the advantage of generally providing a big picture of the design and giving locations of all possible fixed and moving hinge pins. Of course the graphical techniques may be programmed and handled on a computer. The analytic techniques provide greater accuracy and can use previously developed computer programs. Unless care is taken, the analytic techniques could result in very costly and time-consuming trial-and-error procedures.

In both the analytic or graphical techniques, interest is centered on movement of the coupler plane. Since, by the concept of expansion, any link may be considered as a plane and any point in the plane of the link may be considered as a point on the link, the plane containing the coupler link is of prime importance. In the coupler plane there will exist points whose correlated positions as the plane moves generate a circle. Such points are prime candidates for moving hinge pins and the center of the developed circle becomes a fixed hinge pin.

Synthesis techniques may be easily divided into those concerned with finite displacements of the coupler plane and those concerned with infinitesimal displacements of the coupler plane. Techniques coupled with finite displacements are concerned only that the coupler plane move through specified positions and give no consideration to motion of the coupler plane between the specified positions. Generally up to four specified positions are easily considered using finite displacement techniques. Having more than four specified positions presents significant problems in finite displacement synthesis. In addition, velocities and accelerations are not easily included in finite displacement techniques.

Infinitesimal displacement synthesis techniques are more amenable to analytic treatment although many graphical techniques have been developed for infinitesimal displacements. The techniques allow for easy treatment of velocity and acceleration.

6.2   INFLECTION CIRCLE

As a plane which describes a moving link of a mechanism moves, points on the plane describe various paths which differ from one another. In general, the path of points in the plane may pass through inflection points in their path at some instant. At any instant under consideration the locus of all those points which are passing through an inflection point in their paths, or are moving in a straight line, will lie on a circle known as the *inflection circle*. Note that the inflection circle is an instantaneous locus of points and that its size and location will change as the plane (mechanism) moves.

An inflection point on a path is instantaneously a straight line. Thus all points which lie on the inflection circle may be considered as moving in a straight line at the instant under consideration. Points which move in a straight line have a radius of curvature of their path equal to infinity and thus have zero normal acceleration. This fact makes it possible to determine the radius of curvature of the path of any point on the moving plane. The inflection circle is therefore a very useful concept making possible the analysis and design of complex mechanisms.

The instant center for velocity between the coupler link and the fixed link of a mechanism is known as the *pole point*. The pole point is a point of zero velocity on the coupler plane. However, the pole point is not a fixed point in the plane since as the mechanism moves to another configuration, the location of the pole point changes (Fig. 6.1). The locus of the pole point as the mechanism moves considering link 1 as fixed is known as the *fixed polode* and is given the symbol π_f. If the coupler link is held fixed, the locus of the pole point is known as the *moving polode* and given the symbol π_m. The fixed and moving polodes are always in contact at the pole point. As the coupler link moves, the arc distance $P_{(2, 4)}P'_{(2, 4)}$ is equal to the arc distance $P_{(4, 2)}P'_{(4, 2)}$. This implies that the pole point moves along the fixed polode without slipping.

In Fig. 6.2, the fixed and moving polodes are shown in contact at the pole point. Perpendicular to the two polodes at the pole point and extending outward from the fixed polode is the pole normal axis. At 90°

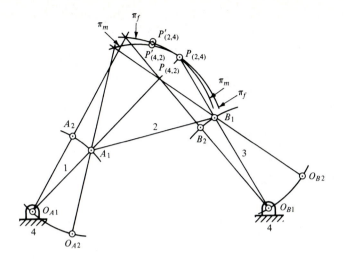

FIGURE 6.1
Fixed and moving polodes.

CCW from the pole normal axis is the pole tangent axis. The center of curvature of the fixed polode is located at point O_f, and the center of curvature of the moving polode is located at point O_m. Distances in Fig. 6.2 are directed distances such that distance $PC = -$distance CP. The point C is an arbitrary point on the moving polode (or on the coupler link). Point C is located a distance r from the pole point in a direction θ from the pole normal axis. As the moving polode moves through a small angle $d\epsilon$ counterclockwise, point O_m moves to position O_m', the pole point moves from position P to position P', and the point C moves from position C_1 to C_2. The center of curvature of the path of point C is located at O_c.

$$\text{Arc distance } C_1 C_2 = PC \, d\epsilon = r \, d\epsilon$$

Also $$\text{Arc distance } C_1 C_2 = O_c C \, d\Omega$$

Then $$O_c C \, d\Omega = r \, d\epsilon$$

or $$d\Omega = \frac{r \, d\epsilon}{O_c C} \tag{6.1}$$

From Fig. 6.2b

$$O_c P \, d\Omega = dP \cos \theta \tag{6.2}$$

Substituting Eq. (6.1) into (6.2) gives

$$\frac{(O_c P) r \, d\epsilon}{O_c C} = dP \cos \theta$$

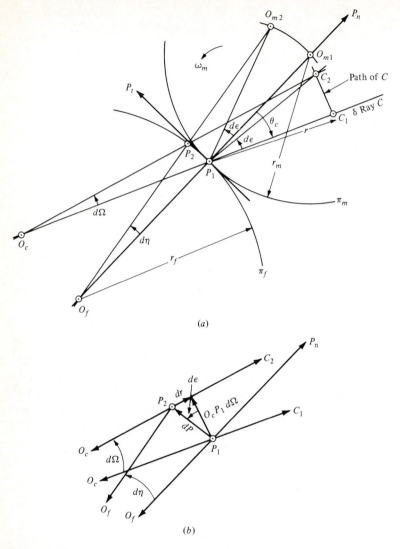

FIGURE 6.2
Geometry associated with fixed and moving polodes.

or

$$\frac{d\epsilon}{dP} = \frac{O_cC \cos \theta}{(O_cP)r} = \text{a constant} = \frac{1}{d} \qquad (6.3)$$

The value of $d\epsilon/dP$ is a constant (called $1/d$) since $d\epsilon$ represents angular displacement of the coupler plane and dP represents the corresponding linear displacement of the pole point. Thus the value of $d\epsilon/dP$ is the same for any and all points on the moving coupler plane.

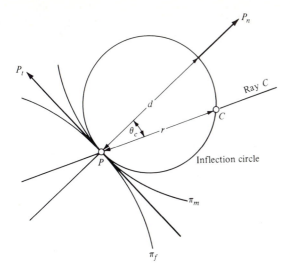

FIGURE 6.3
Inflection circle.

The distance $O_cC = O_cP + PC$ or $O_cC = O_cP + r$ and may be substituted into Eq. (6.3) giving

$$\left(\frac{1}{O_cP} + \frac{1}{r}\right)\cos\theta = \frac{1}{d} \tag{6.4}$$

For all points on the moving plane, the distance from the pole point is r. For those points on the moving plane whose path radius of curvature is infinite, O_cP must be infinite. This includes all points moving in a straight line or passing through an inflection point on their paths. Then

$$\left(\frac{1}{\infty} + \frac{1}{r}\right)\cos\theta = \frac{1}{d}$$

or
$$r = d\cos\theta \tag{6.5}$$

Equation (6.5) is the equation of a circle of diameter d with a chord of length r located at an angle θ from the diameter (the pole normal axis). The circle is known as the inflection circle (Fig. 6.3). The line through the pole point P and the point C (which is moving through an inflection of its path or on a straight line) is known as a *ray*. Equation (6.5) indicates that there is one and only one point on a ray which lies on the inflection circle. Since the point on ray C which lies on the inflection circle is a particularly interesting point, it is given the designation I_c and its distance from the pole point is designated as r_c.

6.3 EULER-SAVARY EQUATION

Leonard Euler (1707–1783) was a very influential Swiss mathematician who spent most of his life outside of Sweden as a professor of mathematics for more than 30 years in Russia and for 25 years in Berlin. At the age of 19 he received a prize for his dissertation on ships' masts from the Paris Academy of Sciences. Euler's contribution to the study of mechanisms lies in his application of algebra and geometry to the study of motion. Euler's "Mechanica sive motus scientia analytice exposita" is quoted as being "the first great work in which analysis is applied to the science of movement."

The Euler-Savary equation provides a means for determination of the radius of curvature of the path of a coupler point. By equating Eqs. (6.4) and (6.5) remembering that Eq. (6.5) relates to coupler points moving through inflections in their path or in straight lines, the first form of the Euler-Savary equation results.

$$\left(\frac{1}{O_cP} + \frac{1}{r}\right)\cos\theta = \frac{\cos\theta}{r_c}$$

$$\frac{1}{r_c} = \frac{1}{O_cP} + \frac{1}{r} \tag{6.6}$$

The first form of the Euler-Savary equation is not particularly handy and may be easily modified into the second form as follows remembering that directions are as indicated in Fig. 6.2:

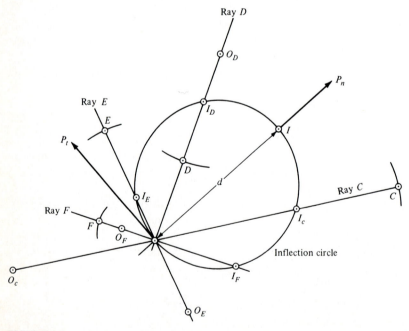

FIGURE 6.4
Application of Euler-Savary equation.

$$r_c = PI_c = PC - I_cC$$

$$O_cP = O_cC - PC$$

$$r = PC$$

Substitution into Eq. (6.6) produces the second form:

$$(PC)^2 = (I_cC)(O_cC) \tag{6.7}$$

Since the left side of Eq. (6.7) is always positive, the two terms on the right side of the equation must be of the same sign. Thus the points I_c and O_c must lie on the same side of point C on ray C.

In Fig. 6.4 the radius of curvature of the path of point D on the coupler plane is found as shown. The centers of curvature of points C, E, and F are also shown.

Point I of Fig. 6.4 is located on the pole normal where it is intersected by the inflection circle. If the point of interest on the coupler plane is located at point I, its path is a straight line. The point is known as Ball's Point.

6.4 BOBILLIER'S CONSTRUCTION

Bobillier's construction is a graphical means of finding the inflection circle based on the second form of the Euler-Savary equation. In Fig. 6.5, point

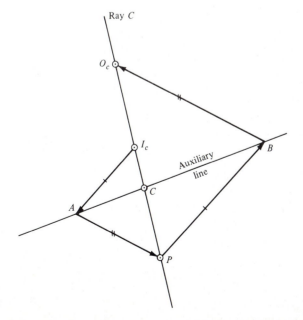

FIGURE 6.5
Bobillier's construction.

P is the pole point through which passes ray C containing point C and point I_c where the inflection circle intersects the ray C. It is desired to locate O_c, the center of curvature of the path of point C, which is also located on ray C.

An auxiliary line is passed in any convenient direction through point C. From point I_c a line is drawn at any convenient angle to intersect the auxiliary line at point A. From point A, a line is drawn to the pole point P. From pole point P a line is drawn parallel to the line I_cA to intersect the auxiliary line at point B. From point B a line is drawn parallel to line AP to intersect ray C at point O_c. With this construction, triangle CI_cA will be similar to triangle CPB. Then

$$\frac{PC}{I_cC} = \frac{BC}{AC} \tag{6.8}$$

Also, triangle CO_cB will be similar to triangle CPA and

$$\frac{O_cC}{PC} = \frac{BC}{AC} \tag{6.9}$$

Equating Eqs. (6.8) and (6.9) results in

$$(PC)^2 = (I_cC)(O_cC) \tag{6.10}$$

which is the second form of the Euler-Savary equation.

The graphical construction may be applied to the four-bar mechanism O_cCDO_d of Fig. 6.6. The coupler link is link 3 and the pole point is instant center $P_{(1,3)}$ located by extending links 2 and 4 to their point of intersection. Ray C passes through the pole point P and coupler point C.

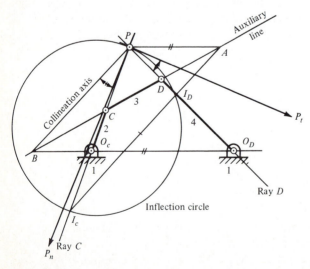

FIGURE 6.6
Bobillier's construction applied to a four-bar mechanism.

Ray C contains the center of curvature of the path of point C or point O_c as well as the point of intersection with the inflection circle, I_c. To locate point I_c an auxiliary line is passed through point C. The auxiliary line may be in any direction, but the direction through point D is most convenient. From point O_c, a line is drawn in any direction to intersect the auxiliary line at point P. Again, the most convenient line will also pass through point O_d. Point B may then be recognized as instant center $P_{(2,4)}$. From point B a line is drawn to the pole point P. This line is known as a collineation axis. From point P a line is drawn parallel to O_cB to intersect the auxiliary line at point A. From point A a line is drawn parallel to the collineation axis to intersect ray C at point I_c. Repeating the process using points D, O_d, and P and ray D will give point I_d. Note that the auxiliary line, points B and A, and the collineation axis have already been determined.

The inflection circle passes through the pole point P and through points I_c and I_d. The diameter of the inflection circle, which is also the pole normal axis, may be found at the intersection of a perpendicular to link 4 at point I_d and a perpendicular to link 2 at point I_c. The pole tangent axis is perpendicular to the pole normal axis and rotated 90° CCW.

Note in Fig. 6.6 that the angle between the pole tangent and ray D is equal to the angle between the collineation axis and ray C. This is Bobillier's theorem.

6.5 HARTMANN'S CONSTRUCTION

Hartmann's technique for locating the inflection circle is a graphical technique based on the definition of the inflection circle diameter as given in Eq. (6.3). Solving Eq. (6.3) for the diameter d and multiplying by $d\tau/d\tau$ where $\tau =$ time gives

$$d = \frac{dP}{d\tau}\frac{d\tau}{d\epsilon}$$

or
$$d = \frac{V_p}{\omega_m} \tag{6.11}$$

where $V_p =$ is the velocity with which the pole point displaces and ω_m is the angular velocity of the moving polode.

In Fig. 6.7, the pole point P is located by extending links 2 and 4 to intersection. The pole point is considered to be a pin sliding in two slots, one slot is on link 2 and the other slot is on link 4. Thus the pole point will always be located at the intersection of extensions of links 2 and 4.

The angular velocity of link 3 is the same as the angular velocity of the moving polode, ω_m. Since point A is a point on link 3, its velocity will be equal to the distance PA multiplied by the angular velocity ω_m. With point A considered as a point on link 2 and all points on link 2 rotating

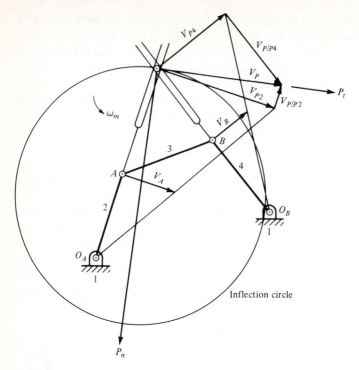

FIGURE 6.7
Hartmann's construction.

about center O_2, the pole point velocity considered as a point on link 2 may be found graphically as V_{P2}. The actual displacement velocity of the pole point will be equal to the velocity V_{P2} plus the sliding velocity of P with respect to link 2 or $V_{P/P2}$ which is parallel to link 2. Then

$$V_P = V_{P2} + V_{P/P2} \qquad (6.12)$$

Applying the same thoughts to link 4,

$$V_P = V_{P4} + V_{P/P4} \qquad (6.13)$$

Simultaneous solution of Eqs. (6.12) and (6.13) will provide V_P, the displacement velocity of the pole point. The pole point is known to displace along the positive pole tangent axis. Therefore, the pole normal axis will be located in a direction rotated 90° from the direction of the pole point displacement velocity in a sense opposite to the angular velocity of the moving polode.

The diameter of the inflection circle may be found by using Eq. (6.11) since the angular velocity of the moving polode and the displacement velocity of the pole point are known.

In using Hartmann's technique for finding the inflection circle it is easiest to assume some convenient value for the angular velocity of the moving polode, ω_m. Once the angular velocity of the moving polode (coupler link) is known, the velocity of points A and B on the coupler may be computed and the graphical procedures follow. It is very convenient to assume the angular velocity of the moving polode is equal to $\frac{1}{2}$. Then the graphical technique will result in the velocity vector V_p which will be equal in length to the radius of the inflection circle. Any assumed value of the moving polode angular velocity will result in the same inflection circle for a given mechanism.

6.6 CUSPIDAL CIRCLE

The inflection circle discussed in Secs. 6.2 through 6.5 deal with the absolute motion of a plane relative to a fixed plane. In some cases it is desirable to deal with the motion of a plane relative to a moving plane. One such instance would be in the determination of the radius of curvature and center of curvature of the path of a point on a moving plane traced on another plane. Such cases are treated as simple kinematic

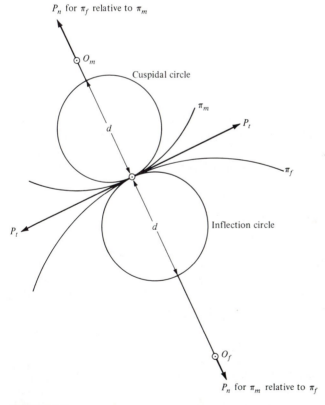

FIGURE 6.8
The cuspidal circle.

inversions of the mechanism in which the reference plane is made stationary. In Fig. 6.8, the inflection circle of the moving polode relative to the fixed polode is found in the usual manner. The cuspidal circle relates motion of the fixed polode relative to the moving polode. The cuspidal circle diameter is the same as the inflection circle diameter. The cuspidal circle is then the inflection circle reflected across the pole tangent axis.

For the mechanism shown in Fig. 6.9, the inflection circle is found using the Bobillier construction. Reflection of the inflection circle over the pole tangent axis produces the cuspidal circle. The inflection circle relates the motion of points on link 3 with respect to link 1. Thus, by application of the Euler-Savary equation and use of the inflection circle, the center of curvature of the path that point C on link 3 will trace on link 1 is located at O_c. The cuspidal circle relates the motion of points on link 1 with respect to link 3. Thus, by application of the Euler-Savary equation and use of the cuspidal circle, the center of curvature of the path that point C' on link 1 will trace on link 3 is located at O'_c.

6.7 CARTER-HALL CIRCLE

The Carter-Hall circle, presented here without proof, is a geometric property of mechanisms discovered by W. J. Carter[*] and expanded by A. S. Hall, Jr.[†]

Each member of a family of four-bar mechanisms designed with the same fixed link (i.e., same fixed pivots) and containing the same velocity ratio of the links (ω_3/ω_1) and the same first and second derivatives of the velocity ratio will have its instant center $P_{(2, 4)}$ located on a circle known as the Carter-Hall circle. The Carter-Hall circle will be centered on the fixed link and pass through instant centers $P_{(2, 4)}$ and $P_{(1, 3)}$ as shown in Fig. 6.10.

6.8 CUBIC OF STATIONARY CURVATURE (CIRCLING POINT CURVE)

The cubic of stationary curvature represents the locus of all points on the moving plane which are moving on curved paths whose radius of curvature is not changing at the particular instant considered. This does not necessarily mean that the path of the point has constant radius of curvature since that radius may be continually changing. It does mean

[*] W. J. Carter, Kinematic Analysis and Synthesis Using Collineation-Axis Equations, *Trans. ASME*, vol. 79, no. 6, pp. 1305–1312, 1957.

[†] A. D. Hall, Jr., Inflection Circle and Polode Curvature, *Trans. Fifth Conf. on Mechanisms*, pp. 207–231, Purdue University, Lafayette, Ind., 1958.

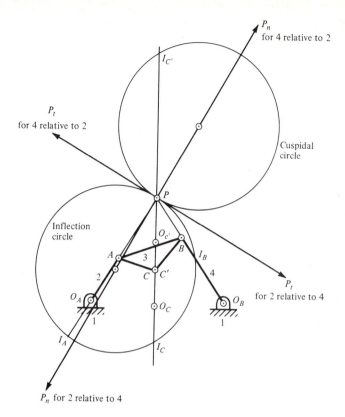

P_n
for 4 relative to 2

$I_{C'}$

P_t
for 4 relative to 2

Cuspidal
circle

P

Inflection
circle

$O_{c'}$

A

B

3

I_B

2

C C'

4

O_A

O_C

O_B

P_t
for 2 relative to 4

1

1

I_A

I_C

P_n for 2 relative to 4

FIGURE 6.9
Use of the cuspidal circle.

that the radius of curvature of the path of the point may have reached a maximum or a minimum. Thus the cubic of stationary curvature is applicable for only the mechanism configuration considered. However, if a point on the moving plane is moving on a circular arc of constant radius, then that point must be located on the cubic of stationary curvature.

If the point C of Fig. 6.11 is in fact a point on the cubic of stationary curvature, the Euler-Savary equation [Eq. (6.7)] provides

$$(PC)^2 = (O_c C)(I_c C)$$

or (remembering that directed distances are considered in the Euler-Savary equation)

$$(PC)^2 = (O_c C)(PC - d_i \cos \theta_c)$$

where d_i is the diameter of the inflection circle. Let $O_c C = R_c$ and $PC = r_c$. Then

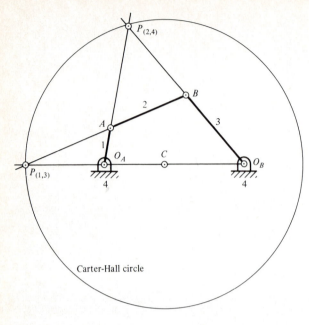

FIGURE 6.10
The Carter-Hall circle.

$$r_c^2 = R_c(r_c - d_i \cos \theta_c)$$

or

$$R_c = \frac{r_c^2}{r_c - d_i \cos \theta_c}$$

If the point C is to have stationary curvature, since it is located on the cubic of stationary curvature, then

$$\frac{dR_c}{dP} = 0$$

where dP is the displacement of the pole point.
Then

$$\frac{dR_c}{dP}$$
$$= \frac{2 + (dr_c/dP)(r + d_i \sin \theta_c) - r_c^2[(dr_c/dP) + d_i \cos \theta_c(d\theta_c/dP) + (dd_i/dP) \sin \theta_c]}{(r_c + d_i \sin \theta_c)^2}$$

From Fig. 6.2b

$$\frac{dr_c}{dP} = -\cos \theta_c$$

$$\frac{d\theta_c}{dP} = \frac{\sin \theta_c}{r_c} - \frac{1}{r_f}$$

where r_f = radius of the curvature of the fixed polode. Since r_f, d_i, and

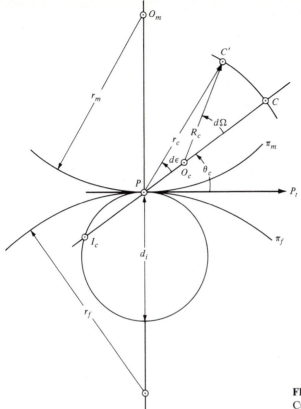

FIGURE 6.11
Cubic of stationary curvature.

dd_i/dP are the same for all points on the moving plane, let

$$\frac{1}{M} = \frac{1}{3}\left(\frac{1}{r_f} - \frac{1}{d_i}\right)$$

$$\frac{1}{N} = -\frac{1}{3}\left(\frac{dd_i/dP}{d_i}\right)$$

Then

$$\frac{dR_c}{dP} = \frac{3r_c^2 d_i \cos\theta_c \sin\theta_c}{(r_c + d_i \sin\theta_c)^2}\left(-\frac{1}{r_c} + \frac{1}{M\sin\theta_c} + \frac{1}{N\cos\theta_c}\right) = 0$$

If dr_c/dP is to become zero, it is sufficient for the quantity in the large parentheses to become zero, and the equation of the cubic of stationary curvature in polar form becomes

$$\frac{1}{r_c} = \frac{1}{M\sin\theta_c} + \frac{1}{N\cos\theta_c}$$

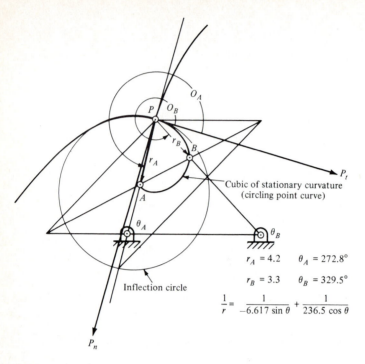

FIGURE 6.12
Construction of circling point curve.

In Fig. 6.12, the points A and B are points on the cubic of stationary curvature and may be used to determine the constants M and N in the equation for the cubic of stationary curvature. Although it is not necessary, the inflection circle is found using the Bobillier construction. This also locates the pole tangent and pole normal axes. The distance PB is called r_b, and its location from the pole tangent axis (CCW) is θ_b. In a similar manner r_a and θ_a are found. With these two values known, the equation of the cubic is solved simultaneously for M and N. Then the cubic of stationary curvature may be plotted. Note that the cubic of stationary curvature is tangent to both the pole normal and pole tangent axes at the pole point.

6.9 GRAPHICAL CONSTRUCTION OF THE CIRCLING POINT CURVE

In Fig. 6.13, the circling point curve is to be constructed for the coupler link of the $O_A ABO_B$ mechanism. Since points A and B must be on the circling point curve, they are used to locate an auxiliary line from which other points on the circling point curve may be located.

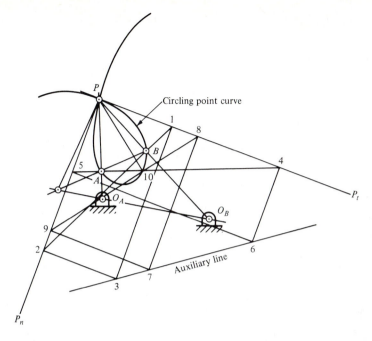

FIGURE 6.13
Graphic construction of circling point curve.

After locating the pole normal and pole tangent axes using the Bobillier construction, a perpendicular to line PB is constructed at B to intersect the pole tangent axis at 1 and the pole normal axis at 2. From point 1 a line is drawn parallel to the pole normal axis. From point 2 a line is drawn parallel to the pole tangent axis. These two lines intersect at point 3. Point 6 is located in a similar manner by constructing a perpendicular to PA at point A and locating points 4 and 5. A line through points 3 and 6 is an auxiliary line. From any point on the auxiliary line (point 7), parallel lines to the pole normal and pole tangent axes locate points 8 and 9. Point 10 on the circling point curve is located such that line 8-9 is perpendicular to line P-10. Selecting any other point on the auxiliary line and following the same procedure will define the circling point curve.

Note that the circling point curve is tangent to the pole tangent axis and is also tangent to the pole normal axis at the pole point. The radius of curvature of the circling point curve as it passes through the pole point tangent to the pole normal axis is the constant N of the equation for the curve. The radius of curvature of the circling point curve as it passes through the pole point tangent to the pole tangent axis is the constant M of the equation for the curve.

Since the equation for the circling point curve is in polar form as derived, substitution of

$$\sin \theta = \frac{y}{r}$$

$$\cos \theta = \frac{x}{r}$$

$$r^2 = x^2 + y^2$$

will convert the equation to cartesian coordinates with the pole normal and pole tangent as coordinates. The equation will become a third-order equation in x and y, and thus the name *cubic of stationary curvature*.

PROBLEMS

6.1. Construct the inflection circle for link 2 related to link 4 and find the center of curvature of the paths of points D, E, and F.

$$O_A O_B = 7.0 \qquad O_A A = 2.0 \qquad O_B B = 4.5 \qquad AB = 4.0$$

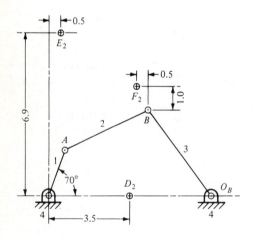

FIGURE P6.1
The inflection circle.

6.2. Locate the inflection circle and find the center of curvature of the path of point C on the coupler.

$$O_A O_B = 7.0 \qquad O_A A = 3.5 \qquad O_B B = 3.0 \qquad AC = CB$$

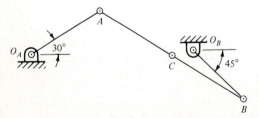

FIGURE P6.2
Centers of curvature.

6.3. Determine the center of curvature O_c of the path of point C on the coupler link. Construct the path of point C and check the curvature by drawing a circle through point O_C.

$$O_A O_B = 3.0 \qquad O_A A = 2.0 \qquad O_B B = 3.5 \qquad AB = 6.5$$
$$BC = 2.0 \qquad AC = 6.0$$

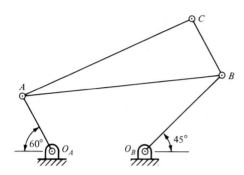

FIGURE P6.3
Coupler point center of curvature.

6.4. Before conducting a velocity or acceleration analysis of the mechanism it is necessary to know the center of curvature of the path of point B. The Hartmann construction is recommended for the inflection circle associated with link 3.

$$BC = 8.0 \qquad AB = 3.5 \qquad O_A A = 1.5 \qquad CD = 4.0$$

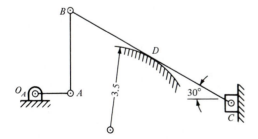

FIGURE P6.4
Hartmann's construction.

6.5. Using Hartmann's construction, locate the inflection circle for the coupler link of the mechanism shown and find the center of curvature of the path of point C.

$$O_A O_B = 10.0 \qquad O_A A = 3.5 \qquad O_B B = 6.0$$
$$AB = 8.5 \qquad AC = 11.5$$

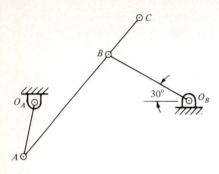

FIGURE 6.5
Hartmann's construction.

6.6. Locate the center of curvature of the path that point C on link 1 would trace on link 3.

$$O_AO_B = 11.0 \qquad O_AA = 2.75 \qquad O_BB = 7.0$$
$$O_AC = 5.0 \qquad AC = 3.0$$

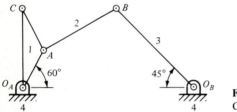

FIGURE P6.6
Cuspidal circle.

6.7. Write a basic program to provide the polar coordinates of the cubic of stationary curvature equation. The constants M and N in the equation $1/R = 1/M \sin \theta + 1/N \cos \theta$ are assumed to be known.

6.8. Plot the cubic of stationary curvature associated with link 2 of the mechanism.

$$O_AO_B = 6.0 \qquad O_AA = 4.0 \qquad O_BB = 3.75$$

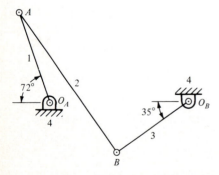

FIGURE P6.8
Cubic of stationary curvature.

6.9. Construct the cubic of stationary curvature associated with link 2. Select the point C on the cubic and find its center of curvature. Lay out the path of point C on link 2 and check its radius of curvature graphically.

$$O_A O_B = 5.0 \qquad O_A A = 2.0 \qquad O_B B = 4.0 \qquad \text{angle } O_A O_B B = 90°$$

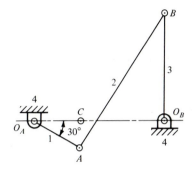

FIGURE P6.9
Cubic of stationary curvature.

CHAPTER
7

GRAPHICAL SYNTHESIS

7.1 INTRODUCTION

Graphical techniques of mechanism synthesis are fast and provide a wealth of information with very little effort. Accuracy of the graphical techniques is dependent upon the care used in conducting the technique. The techniques generally require determination of perpendicular bisectors and intersections of straight lines. Greatest error is introduced in making the scale of the drawing too small or in determination of the point of intersection of two lines at small angles to each other. The accuracy of graphical techniques often results in mechanism errors less than those due to manufacturing tolerances.

The chief advantage of the graphical techniques is that they will provide a global picture of the mechanism and possible pivot locations. In the event that the resulting mechanism is not satisfactory for some reason, it is very easy to modify the design to produce a more satisfactory mechanism. The chief disadvantage of the graphical techniques is that although they will result in a mechanism design which will do the required task, there is no guarantee that the mechanism will be a good mechanism in light of dynamic balance, poor transmission angles, etc. The designer must critically review the design. This is true regardless of the technique used to derive the design.

Mechanism synthesis is concerned with

1. Coordination of input and output crank rotations better known as function generators
2. Generation of a path with or without coordination of input crank positions
3. Coordination of input and output crank angular velocities
4. Specification of transmission angles
5. Some combination of 1, 2, 3, and/or 4.

In the case of function generation, several scale factors must be selected and limits on crank and rocker rotation angles must be decided upon. In general, crank and rocker rotation angles should be limited to less than 120° if possible.

Scale factors relate the angular displacement of the input crank to the independent variable of the function to be generated and the angular displacement of the output rocker to the dependent variable of the function to be generated. Thus, in mechanization of the function $y = \sin x$, the input crank rotation θ is related to the value of x and the output rocker rotation ϕ is related to the value of y. The scaling factors become

$$k_\theta = \frac{\delta\theta}{\delta x} \text{ degrees of crank rotation per unit } x$$

$$k_\phi = \frac{\delta\phi}{\delta y} \text{ degrees of rocker rotation per unit } y$$

7.2 SPECIFIED TRANSMISSION ANGLE

A crank rocker mechanism in position for maximum or minimum transmission angle is shown in Fig. 7.1. Note that the maximum or minimum transmission angle occurs when the input crank is collinear with the fixed link.

To design for a specified transmission angle, the output rocker is first drawn in its two positions as in Fig. 7.2. The line b_1 is constructed at the minimum transmission angle from rocker $O_B B_1$, and the line b_2 is constructed at the maximum transmission angle from crank $O_B B_2$. The arbitrary distance $B_1 a_1'$ is laid off from B_1 along line b_1 to locate point a_1'. The same distance is laid off along line b_2 from point B_2 to locate point a_2'. A straight line through points a_1' and a_2' must intersect point O_B. If this is not the case, another arbitrary distance $B_1 a$ is selected and the process repeated. The procedure is very rapid, and a correct solution is soon achieved.

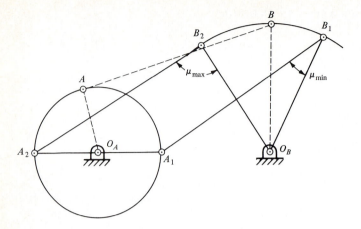

FIGURE 7.1
Limits of transmission angle.

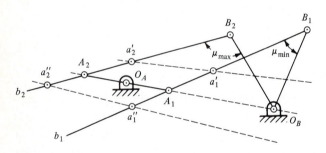

FIGURE 7.2
Design for specified transmission angle.

7.3 CRANK ROCKER MECHANISMS

The crank rocker mechanism is inherently a quick-return mechanism and as such presents many possibilities for its design. Figure 7.3 shows a crank rocker mechanism in its two extreme positions. Extreme positions occur when the input crank and the coupler links are in alignment. The input crank rotation θ_{12} is noted to be larger by the amount α than the input crank rotation θ_{21} which provides the quick-return feature of the crank rocker mechanism known as the transmission ratio TR.

$$\mathrm{TR} = \frac{180}{180 - \alpha}$$

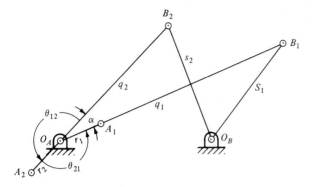

FIGURE 7.3
Crank rocker extreme positions.

If the input crank is rotating at a constant angular velocity, the output rocker will oscillate slower from position B_1 to position B_2 and faster from position B_2 to position B_1.

The distance $O_A B_1$ is equal to the sum of the lengths of the coupler and the crank. The distance $O_A B_2$ is equal to the length of the coupler minus the length of the crank.

$$O_A B_1 = q + r \qquad O_A B_2 = q - r \qquad (7.1)$$

Simultaneous solution of Eqs. (7.1) provides

$$q = \frac{O_A B_1 + O_A B_2}{2} \qquad r = \frac{O_A B_1 - O_A B_2}{2} \qquad (7.2)$$

A mechanism may be designed for a specified transmission ratio and output rocker displacement by constructing the output rocker in its two extreme positions using any convenient scale as in Fig. 7.4. On the perpendicular bisector of $B_1 B_2$, the point O is located so that the angle $B_1 O B_2$ is equal to 2α. Point O is the center of a circle through B_1 and B_2. The fixed pivot O_A must lie on this circle since the angle $B_1 O_A B_2$ will be equal to the angle α. Selection of a site for O_A on the circle and application of Eqs. (7.2) allows specification of the entire mechanism.

Naturally the designer must exercise some thought in selecting the site for fixed pivot O_A. If O_A is selected on the perpendicular bisector of $B_1 B_2$, the mechanism will be mathematically correct but will be one which has extremely poor transmission angles and will probably not operate at all.

For the special case of a specified transmission ratio and also a specified initial position of the input crank θ_1, the fixed pivot O_A must be located on the circle centered at O at the intersection with another circle centered at P. Point P is located on the perpendicular bisector of $O_B B_1$

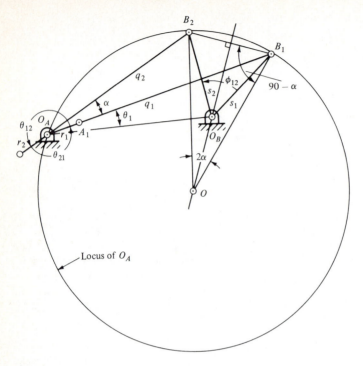

FIGURE 7.4
Quick-return mechanism with specified transmission ratio.

such that the angle B_1PO_B is equal to $2\theta_1$. However, the magnitude of θ_1 cannot be a free choice. If α is less than output crank displacement ϕ_{12}, the fixed pivot O_B is located inside the circle containing pivot O_A, and θ_1 must be less than $90° - \frac{1}{2}\phi_{12}$. If α is greater than output crank displacement ϕ_{12}, the fixed pivot O_B is located outside the circle containing pivot O_A, and θ_1 must be greater than $90° - \frac{1}{2}\phi_{12}$. In the case that $\alpha = \phi_{12}$, a four-bar mechanism is possible only if $\theta_1 = \frac{1}{2}\phi_{12}$ and then an infinite number of mechanisms are possible.

7.4 INVERSION

Inversion is a very powerful technique for graphical synthesis of mechanisms. Inversion is a process by which a link which is normally fixed is allowed to move and another link which normally moves is held stationary. The technique is based on the assumed condition that the length of links does not change during motion. Inversion will not result in any alteration of relative motions of the links in a mechanism.

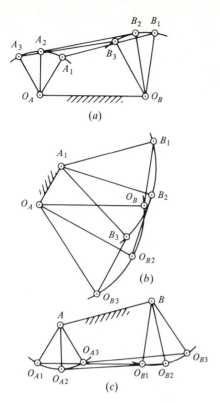

(a)

(b)

(c)

FIGURE 7.5
Inversion.

A four-bar crank rocker mechanism is shown in three positions in Fig. 7.5a. If the mechanism is inverted by allowing the fixed link p to move while the input crank q is held stationary, the three positions become as shown in Fig. 7.5b. The inversion is accomplished by noting that the length of p does not change. In this case the locus of O_B becomes a circle centered at O_A. The positions of O_B on its locus are found by laying off the distance A_2O_B from A_1 to locate O_{B2} and laying off the distance A_3O_B from A_1. The inverted positions of B must lie on a circle centered at A_1, and the distance from O_B to B is constant. Therefore, $O_BB_1 = O_{B2}B_2 = O_{B3}B_3$. This described procedure results in inverting the mechanism to position 1. If arcs and distances were to be laid off from A_2 in place of A_1, the mechanism would be inverted into position 2. Inversion with the coupler link 2 fixed in position 1 is shown in Fig. 7.5c.

7.5   ROTATION POLE POINTS

The coupler plane of a mechanism is shown in two positions described by the position of line AB in Fig. 7.6. The perpendicular bisector of positions A_1A_2 and the perpendicular bisector of positions B_1B_2 will

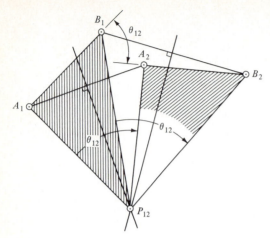

FIGURE 7.6
Rotation pole point.

interact at the rotation pole point P_{12}. In order to move the coupler plane (line AB) from position 1 to position 2 it is simply necessary to rotate the plane about center P_{12} through the rotation angle θ_{12}. Since the distances $A_1P_{12} = A_2P_{12}$, $B_1P_{12} = B_2P_{12}$, and $A_1B_1 = A_2B_2$, the triangle $A_1P_{12}B_1$ is congruent to the triangle $A_2P_{12}B_2$.

In the event that it is not convenient to use the pole point as a pivot point, an infinite number of four-bar mechanisms to move the coupler plane from position 1 to position 2 are possible. The fixed pivot of point A may be located anywhere on the perpendicular bisector of A_1A_2, and the fixed pivot of point B may be located anywhere on the perpendicular bisector of B_1B_2 as shown in Fig. 7.7. On inspection of Fig. 7.7 it is noted that the distance A_1B_1 and the distance A_2B_2 are both "seen" by the pole point P_{12} at the angle β. At the same time, the fixed link distance O_AO_B is also seen by the pole point at the angle β. Therefore, the fixed link and the coupler link of a four-bar mechanism moving through two positions subtend equal angles at the pole point.

In the event that it is not convenient to use the points A and B as moving hinge pins or to place the fixed hinge pins on the perpendicular bisectors of A_1A_2 and B_1B_2, other sites may be selected arbitrarily. In Fig. 7.8, the fixed hinge pin O_c and the moving hinge pin D are selected as more desirable hinge pins. At the same time, the input crank O_cC is to rotate through the angle ϕ_{12} and the output crank is to rotate through the angle μ_{12} while the moving plane rotates through the angle θ_{12}. In order to locate the moving hinge pin C on the coupler plane, the line $P_{12}C_1$ is laid off at an angle $-\frac{1}{2}\theta_{12}$ from the line O_cP_{12}. The line O_cC_1 is laid off at an angle $-\frac{1}{2}\phi_{12}$ from the line O_cP_{12}. In order to locate the fixed hinge pin O_D on the fixed plane, the line O_DP_{12} is laid off at an angle $+\frac{1}{2}\theta_{12}$ from the line DP_{12}. The line O_DD is laid off at the angle $-\frac{1}{2}\mu_{12}$ from the line

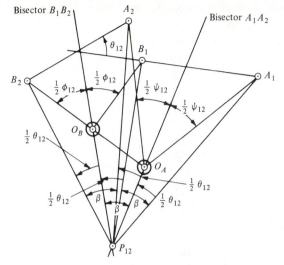

FIGURE 7.7
Fixed centers on bisectors.

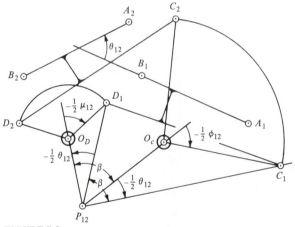

FIGURE 7.8
Arbitrary location of fixed and moving hinge pins.

$O_D P_{12}$. Now as point C rotates through the angle θ_{12} the output crank will rotate through the angle ϕ_{12} and the coupler plane will move from position 1 to position 2. Note that the angle $C_1 P_{12} D_1 =$ angle $O_c P_{12} OD$.

7.6 COUPLER CURVE GENERATION

Use of a pole point for a fixed or for a moving hinge pin makes it possible to design a mechanism which will trace a specified coupler curve with accuracy at as many as six points. Accuracy at three points on a coupler curve is a trivial problem. Accuracy at five points is discussed, and the technique may be adopted to four-point accuracy very easily. Accuracy at six points involves some trial and error.

The technique may be applied to use of the pole point as a moving or as a fixed hinge pin. In Fig. 7.9 the pole point is used as a fixed hinge pin. It is required to design a mechanism which will trace a curve through points C_1 to C_5. The technique requires that two pole points be located at the same site. In selecting the pole points to be used, one set of points must completely enclose the other set. Therefore the following sets of pole points may be used: P_{15}, P_{24}; P_{15}, P_{23}; P_{15}, P_{34}; P_{14}, P_{23}; P_{25}, P_{34}. For this example pole points P_{15}, P_{34} are selected. Pole point P_{15} must be located on the perpendicular bisector of positions C_1 and C_5. Pole point P_{34} must be located on the perpendicular bisector of positions C_3 and C_4. The two perpendicular bisectors will intersect, and the pole points P_{15} and P_{34} are both located at the point of intersection. Fixed pivot O_B is located at the two coincident pole points.

The other fixed pivot, O_A, is located along the fixed link at any convenient direction from pivot O_B. The rays a_1O_B and a_5O_B are

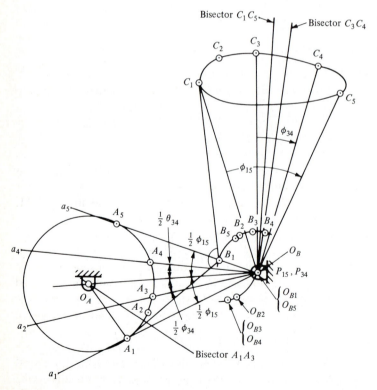

FIGURE 7.9
Path generation.

constructed at the angle $\frac{1}{2}\phi_{15}$ from the fixed link. The rays a_3O_B and a_4O_B are constructed at the angle $\frac{1}{2}\phi_{34}$ from the fixed link. By selection of a convenient length CA, the point A_1 is located on ray a_1 by striking an arc from point C_1. In the same manner point A_3 is found on ray a_3 by striking the same arc from point C_3. A perpendicular bisector of points A_1 and A_3 will intersect the fixed link at point O_A. Points A_2, A_4, and A_5 are located on an arc through point A_1 centered at O_A by striking arcs from points C_2, C_4, and C_5.

Inverted positions of O_B are found by constructing congruent triangles.

Triangle $A_1O_{B2}C_1$ is congruent to triangle $A_2O_BC_2$.
Triangle $A_1O_{B3}C_1$ is congruent to triangle $A_3O_BC_3$.
Triangle $A_1O_{B4}C_1$ is congruent to triangle $A_4O_BC_4$.
Triangle $A_1O_{B5}C_1$ is congruent to triangle $A_4O_BC_5$.

The moving hinge pin B in position 1 is located at the center of the circle through the inverted positions of O_B.

7.7   POLE TRIANGLE

With three positions of a coupler plane, three pole points are identified. Pole point P_{12} is located at the intersection of perpendicular bisectors of A_1A_2 and B_1B_2. Using points A_2A_3 and B_2B_3 locates pole point P_{23}. Points A_1A_3 and B_1B_3 locate pole point P_{13}. The three pole points may be connected to form a pole triangle as shown in Fig. 7.10.

The sides of the pole triangle are numbered in accordance with their endpoint numbers. Side 1 has endpoints P_{12} and P_{13}. The subscript 1 is common to both endpoints, thus the line joining the endpoints is named side 1. In a similar manner, sides 2 and 3 are numbered as in Fig. 7.11. Pole point P_{12} represents a center of rotation of the coupler plane from position 1 to position 2. The pole point P_{12} is thus a point on the coupler plane that does not change position on the fixed plane as the coupler plane moves from position 1 to position 2. Pole point P_{23} is a point on the moving plane in position 2 that does not change position on the fixed plane as the coupler moves from position 2 to position 3. To locate the position of pole point P_{23} on the moving plane when the plane is in position 1, it is necessary to use pole point P_{12} to rotate pole point P_{23} backwards through the angle ϕ_{12}. Since the plane moves from position 1 to position 2 by rotation about pole point P_{12}, the arc of rotation centered at pole point P_{12} and passing through pole point P_{23} represents the path of the pole point when the plane moves from position 1 to position 2. Since the pole point P_{23} represents a point on the moving plane in

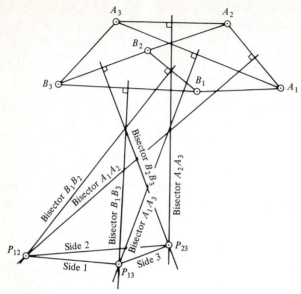

FIGURE 7.10
The pole triangle.

position 3, the pole point P_{13} may be used as a center of rotation to rotate pole point P_{23} backwards to position 1 of the moving plane. Thus, point P'_{23} represents the position of pole point P_{23} on the coupler plane when the moving plane is in position 1.

In locating the position of P_{23} in position 1 (phase 1) of the moving plane, it was rotated about pole point P_{13} through the angle $-\phi_{13}$. It was also rotated about pole point P_{12} through the angle $-\phi_{12}$. With this construction, it is seen that the internal angle between side 1 and side 2 of the pole triangle is equal to $\frac{1}{2}\phi_{12}$ and that its direction is such that side 1 of the triangle would rotate about P_{12} into side 2 of the triangle. The

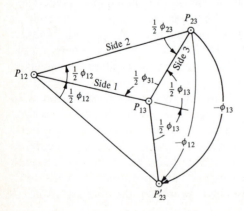

FIGURE 7.11
Sides of the pole triangle.

internal angle of the pole triangle at P_{23} is equal to $\frac{1}{2}\phi_{23}$ and the internal angle of the pole triangle at P_{13} is equal to $\frac{1}{2}\phi_{31}$. Thus the external angle of the pole triangle at P_{13} (between sides 3 and 1) is equal to the angle $\frac{1}{2}\phi_{13}$.

7.8 USE OF THE POLE TRIANGLE

For three finite positions of a coupler plane, the pole triangle provides all information necessary to describe motion of the plane. In Fig. 7.12 the pole triangle for three positions of a moving plane is given with the displacement angles indicated. Point C on the moving plane in position 1 is given, and it is desired to locate point C on the fixed plane when the moving plane has moved to positions 2 and 3. Using pole point P_{12}, the point C is rotated in the indicated direction through the angle ϕ_{12} to locate point C in position 2. Using pole point P_{23}, the point C may be rotated (from its position in phase 2) in the indicated direction through the angle ϕ_{23} to locate point C in position 3 (phase 3). As a check on the procedure, using pole point P_{13} and rotating point C in phase 3 through the angle ϕ_{31} in the indicated direction will return point C to its original position in phase 1.

A construction not requiring measurements of angles is possible as shown in Fig. 7.13. The pole triangle is shown with point C_1 in phase 1,

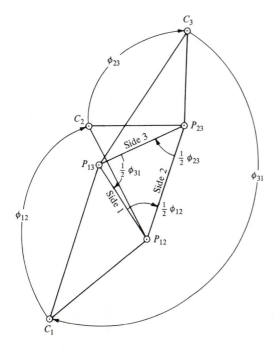

FIGURE 7.12
Rotation using pole points.

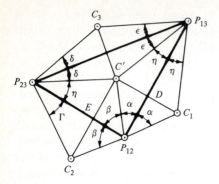

FIGURE 7.13
Cardinal point of pole center triangle.

and it is desired to locate point C in phases 2 and 3 of the motion. The position of point C_1 is reflected across side one of the pole triangle to locate point C' (known as a *cardinal* point). Simple reflection of the cardinal point across side 2 of the pole triangle will locate point C in phase 2. Simple reflection of the cardinal point across side 3 of the pole triangle will locate point C in phase 3 of the motion. Reflection of point C_1 across side 1 to locate cardinal point C' means that $C_1 C'$ is perpendicular to side 1 and that distance $C_1 D$ is equal to distance DC'. Therefore, the angle $\alpha = C_1 P_{12} D$ is equal to the angle $\alpha = DP_{12} C'$. The line $C'C_2$ is perpendicular to side 2 of the triangle, and distance $C'E$ is equal to distance EC_2. Therefore, the angle $\beta = C'P_{12}E$ is equal to the angle $\beta = EP_{12}C_2$. The angle $DP_{12}E = (\alpha + \beta)$ is the internal angle of the pole triangle and is equal to one-half the angular displacement of the moving plane from phase 1 to phase 2. Since the angle $C_1 P_{12} C_2$ is equal to $2\alpha + 2\beta$, it represents the total rotational angle from phase 1 to phase 2 of the moving plane. In like manner, position C_3 may be located using C'.

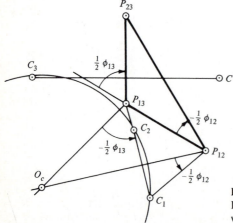

FIGURE 7.14
Locating a point moving on a circle with designated center.

If the pole center triangle for three positions of a coupler plane is known and it is desired to locate a point C which moves in a circle centered at O_C, the procedure is as shown in Fig. 7.14. The line $P_{12}C_1$ is laid off from the line $O_C P_{12}$ at an angle equal to $-\frac{1}{2}\phi_{12}$ which is determined from the pole center triangle. The line $P_{13}C_1$ is laid off at an angle equal to $-\frac{1}{2}\phi_{13}$ from the line $O_C P_{13}$. In its three correlated positions the points C_1, C_2, and C_3 will all lie on a circle centered at O_C. The reverse procedure may be used to locate the center of a circle containing the three correlated positions of any point on the moving plane.

7.9 ORTHOCENTER OF THE POLE CENTER TRIANGLE

The *orthocenter* of the pole triangle shown as point O in Fig. 7.15a is located at the intersection of lines drawn from the pole points perpendicular to opposite sides of the triangle. It is not necessary for the orthocenter to lie inside the triangle as shown in Fig. 7.15b.

If the orthocenter of a pole triangle is used as a cardinal point and is reflected across the sides of the pole triangle, points O_1, O_2, and O_3 are located as shown in Fig. 7.16.

In the right triangle $P_{12}aP_{23}$ of Fig. 7.16, the sum of the angles α, β, Γ must be 90°. Then in the isosceles triangle $O_2P_{12}O_1$, the angle $P_{12}O_2O_1$ must equal Γ.

Since the line CP_{13} is perpendicular to the line $P_{12}P_{23}$ and the line $P_{12}P_{13}$ is perpendicular to the line aP_{23}, the angle $CP_{13}P_{12}$ is equal to Γ. Consequently the angle $P_{12}P_{13}O_1$ is also equal to Γ.

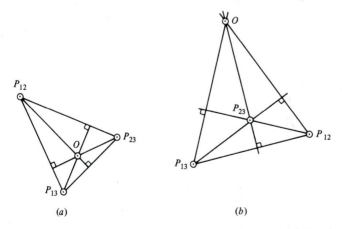

(a)

(b)

FIGURE 7.15
Orthocenter of pole triangle.

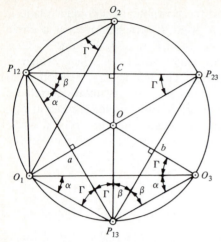

FIGURE 7.16
Circumscribing circle of pole triangle.

Since the line CP_{13} is perpendicular to the line $P_{12}P_{23}$ and the line $P_{12}b$ is perpendicular to the line $P_{23}P_{13}$, the angle $CP_{13}P_{23}$ is equal to β. Consequently the angle $P_{23}P_{13}O_3$ is also equal to β. Then in the isosceles triangle $O_3P_{13}O_1$ the angle $P_{13}O_3O_1$ must be equal to α.

In the right triangle $P_{13}bO_3$, the angle $P_{13}O_3b$ must be equal to $90 - \beta$ or $\alpha + \Gamma$. Therefore, the angle $P_{12}O_3O_1$ must equal Γ. The line segment $P_{12}O_1$ is seen by the angle Γ from points O_2, P_{23}, O_3, and P_{13}. The only way this is possible is if all six points representing the poles of the pole center triangle and reflected positions of the orthocenter all lie on the same circle. The circle is known as the *circumscribing* circle of the pole center triangle.

If the circumscribing circle of the pole center triangle is reflected across each side of the pole center triangle, the three circles (known as image circles) will intersect at the orthocenter of the pole center triangle as shown in Fig. 7.17.

In the event that a point is to have its three correlated positions on a straight line, the straight line must pass through the orthocenter of the pole center triangle, and each of the three correlated positions of the point will lie on its respective image circle.

Figure 7.18 shows the pole center triangle, the three correlated positions of point C, and the orthocenter O of the pole center triangle. If the three positions of point C are to be on a straight line the perpendicular bisectors of C_1C_2, C_1C_3, and C_2C_3 must be parallel and perpendicular to the straight line containing C_1, C_2, and C_3.

At C_1, the angle $P_{12}C_1P_{13} = \frac{1}{2}\phi_{13} - \frac{1}{2}\phi_{12} = \frac{1}{2}\phi_{23}$.

Since $P_{13}O$ is perpendicular to $P_{12}P_{23}$ and $P_{12}O$ is perpendicular to $P_{23}P_{13}$,

$$\text{Angle } P_{12}OP_{13} = \text{angle } P_{12}P_{23}P_{13} = \tfrac{1}{2}\phi_{23}$$

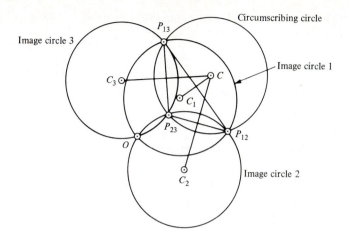

FIGURE 7.17
Pole triangle image circles.

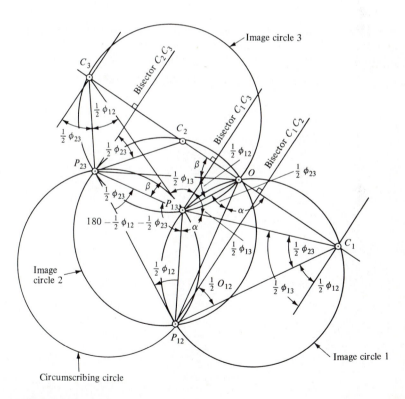

FIGURE 7.18
Three positions on a straight line.

Since $P_{12}P_{13}$ is seen by equal angles at C_1 and at O, the points C_1, O, P_{12}, and P_{13} must all lie on a circle—image circle 1.

At C_3, the angle $P_{23}C_3P_{13} = \frac{1}{2}\phi_{12}$.

Since $P_{23}O$ is perpendicular to $P_{12}P_{23}$ and $P_{12}O$ is perpendicular to $P_{23}P_{13}$,

$$\text{Angle } P_{23}C_3P_{13} = \text{angle } P_{23}OP_{13} = \tfrac{1}{2}\phi_{12}$$

Since $P_{23}P_{13}$ is seen by equal angles at C_3 and at O, the points C_3, O, P_{23}, and P_{13} must all lie on a circle—image circle 3.

At C_2

$$\text{Angle } P_{23}OP_{12} = \text{angle } P_{23}OP_{13} + \text{angle } P_{13}OP_{12}$$

$$\text{Angle } P_{23}OP_{12} = \tfrac{1}{2}\phi_{23} + \tfrac{1}{2}\phi_{12} = \tfrac{1}{2}\phi_{13}$$

Since $P_{23}P_{13}$ is perpendicular to $P_{12}O$ and since $P_{13}O$ is perpendicular to $P_{12}P_{23}$,

$$\text{Angle } P_{23}OP_{12} = \text{angle } \tfrac{1}{2}\phi_{13} = 180 - \tfrac{1}{2}\phi_{12} - \tfrac{1}{2}\phi_{23}$$

Therefore, points C_2 and O see the line $P_{12}P_{23}$ at the same angle and points C_2, O, P_{12}, and P_{23} must lie on a circle—image circle 2.

On image circle 1, angle $P_{12}P_{13}C_1$ must be equal to angle $C_1OP_{12} = \alpha$. On image circle 3, angle C_3OP_{23} must be equal to angle $C_3P_{13}P_{23} = \beta$.

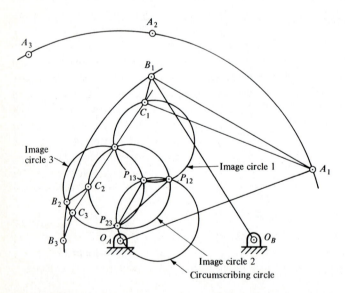

FIGURE 7.19
Locating a coupler point with three positions on a straight line.

At P_{13} the sum of the angles becomes

$$(180 - \tfrac{1}{2}\phi_{12} - \tfrac{1}{2}\phi_{23}) + \alpha + \tfrac{1}{2}\phi_{13} + \tfrac{1}{2}\phi_{13} + \beta = 360$$

$$\alpha + \tfrac{1}{2}\phi_{13} + \beta = 180$$

At O the angle $C_1 O C_3 = \alpha + \tfrac{1}{2}\phi_{13} + \beta = 180$ or the line containing C_1, C_2, and C_3 must be a straight line passing through point O.

In Fig. 7.19 a four-bar mechanism is shown in three correlated positions. It is desired to locate a coupler point C such that in the three positions of the mechanism that coupler point will lie on a straight line. The pole center triangle, its circumscribing circle, and three image circles which intersect at the orthocenter of the pole center triangle are shown.

A straight line through the orthocenter of the triangle at any angle will intersect the image circles at points C_1, C_2, and C_3 indicating the three positions of coupler point C.

7.10 RELATIVE POLE POINTS

The problem of coordinating crank and rocker displacements is more easily handled by use of the relative pole points. The principle of inversion is utilized in locating relative pole points which are given the letter R to distinguish them from pole points. In the mechanism of Fig. 7.20, the input crank is to rotate from position 1 to position 2 through the displacement angle θ_{12} and from position 2 to position 3 through the displacement angle θ_{23}. At the same time the output rocker is to rotate from position 1 to position 2 through the displacement angle ϕ_{12} and from position 2 to position 3 through the displacement angle ϕ_{23}. Such a requirement is known as a *function generator*.

In order to locate relative pole 12, the mechanism is inverted from the second position to the first position. The point A would normally displace to point A_2. In the inverted position the input crank is held stationary and the remainder of the mechanism is allowed to move. Point O_B will rotate about center O_A through the angle $-\theta_{12}$ and be located at O_B'. The distance $O_B B$ remains constant and the distance AB remains constant so that point B is inverted to position B'. Perpendicular bisectors of $O_B O_B'$ and BB' intersect to locate relative pole R_{12}. Note that the angle between $O_A O_B$ and the perpendicuar bisector of $O_B O_B'$ is equal to $-\tfrac{1}{2}\theta_{12}$.

With the output rocker held stationary in position 1, the remainder of the mechanism must rotate through the angle $-\phi_{12}$. The point O_A rotates about center O_B through the angle $-\phi_{12}$ and is located at O_A'. Note that the angle between $O_A O_B$ and the perpendicular bisector of $O_A O_A'$ is equal to $-\tfrac{1}{2}\phi_{12}$.

Therefore, to locate the relative poles it is only necessary to construct negative half angles from the originally fixed link. As shown in

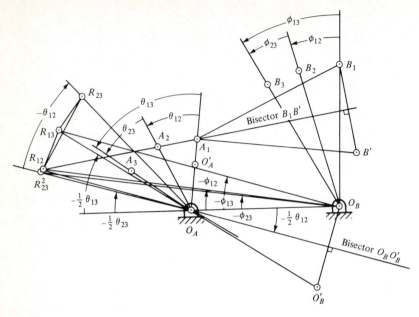

FIGURE 7.20
Relative pole triangle.

Fig. 7.20, the relative poles R_{12}, R_{13}, and R_{23} are located by constructing negative half angles. However, relative poles R_{12} and R_{13} have been located by inversion to position 1 of the mechanism, while relative pole R_{23} is located by inversion to position 2 of the mechanism. In order to locate relative pole R_{23} as inverted to position 1 it is necessary to rotate pole R_{23} from position 2 to position 1 through the angle $-\theta_{12}$. In this manner the relative pole center triangle relating to position 1 of the mechanism can be drawn. The relative pole center triangle possesses all the same properties as the pole center triangle.

Figure. 7.21 shows a slider crank mechanism for which the relative pole triangle relating motion of the slider to the crank is desired. Since the slider moves in a straight line, its center of rotation is located at infinity and the line joining the two centers of rotation is a vertical line. Displacement of the slider is in terms of linear motion rather than angular displacement. Starting at point O_A the distance $-\frac{1}{2}l_{12}$ is laid off to intersect the ray at an angle of $-\frac{1}{2}\theta_{12}$ from the line of centers. In this manner relative pole R_{12} is located. Laying off distance $-\frac{1}{2}l_{23}$ and angle $-\frac{1}{2}\theta_{23}$ locates relative pole R_{23} in its second position. Rotation of the relative pole R_{23} through the angle $-\theta_{12}$ locates the pole in position 1. Laying off distance $-\frac{1}{2}l_{13}$ and angle $-\frac{1}{2}\theta_{13}$ locates relative pole R_{13}, and the relative pole triangle may be drawn as shown.

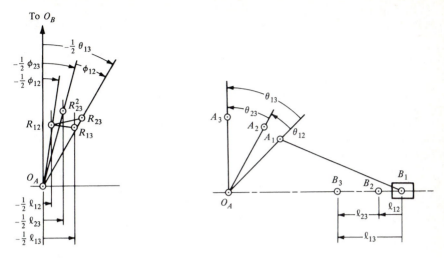

FIGURE 7.21
Relative pole triangle for a slider crank mechanism.

7.11 BURMESTER'S CURVE

In the case of four positions of the moving plane it is possible to identify six pole points and to locate points on the moving plane such that their four positions lie on a circle. There is an infinity of such points whose locus lies on a curve known as Burmester's curve. Ludwig Burmester was a professor of kinematics and geometry at Munich. He wrote on many subjects which were amenable to geometric description. His *Lehrbuch der Kinematik* published in 1888 is a very systematic and thorough treatment of theoretical kinematics. The book remains as a classic today. *Lehrbuch der Kinematik* is actually two volumes. Volume 1 contains only text (in German) and volume 2 contains all the required illustrations.

If the four positions of a point on the moving plane are to lie on a circle as shown in Fig. 7.22, the pole points may be located by constructing perpendicular bisectors. The pole point P_{12} must be located on the perpendicular bisector of $A_1 A_2$ so that wherever the pole point P_{12} is located it is somewhere along the line a_{12}. In a similar manner the lines passing through pole points P_{13}, P_{14}, P_{23}, P_{24}, and P_{34} may be located. Since the four positions of point A lie on a circle, the chord $A_1 A_2$ is seen by points A_3 and A_4 through the angle α. The line a_{23} is perpendicular to chord $A_2 A_3$, and the line a_{13} is perpendicular to chord $A_1 A_3$; therefore, the angle between line a_{23} and line a_{13} is also equal to the angle α. In a similar manner the angles β, Γ, and ϵ are identified.

If the point O_A is to serve as the center of the circle containing the four positions of point A, the point O_A must see the line joining pole

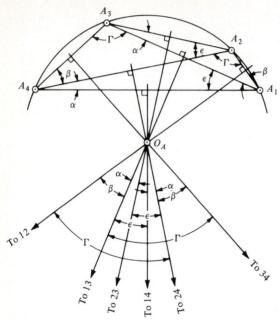

FIGURE 7.22
Pole points for four positions on a circle.

points P_{13} and P_{23} and the line joining pole points P_{14} and P_{24} through the same angle α. The four pole points form a pole center quadrilateral known as an opposite pole quadrilateral since the set of pole points P_{13} and P_{24} and the set P_{14} and P_{23} do not contain common subscripts. Several pole center quadrilaterals are possible by selecting pairs of pole points which do not contain common subscripts. Those sides of the quadrilateral formed by connecting pole points with common subscripts are known as the *opposite sides*. The opposite sides of a pole center quadrilateral see points which serve as the center of circles through four positions of a point on a coupler through the same angle (or angles which differ by 180°). Those pole point sets which may be used to construct Burmester's curve include

Common subscripts	Pole point sets
1 and 2	$P_{13}P_{23}, \; P_{14}P_{24}$
1 and 3	$P_{12}P_{13}, \; P_{14}P_{34}$
1 and 4	$P_{12}P_{24}, \; P_{13}P_{34}$
2 and 3	$P_{12}P_{13}, \; P_{24}P_{34}$
2 and 4	$P_{12}P_{14}, \; P_{23}P_{34}$
3 and 4	$P_{13}P_{14}, \; P_{23}P_{24}$

In Fig. 7.23a, the four positions of a crank are shown. It is desired to locate the rocker of the mechanism such that a coupler point will have

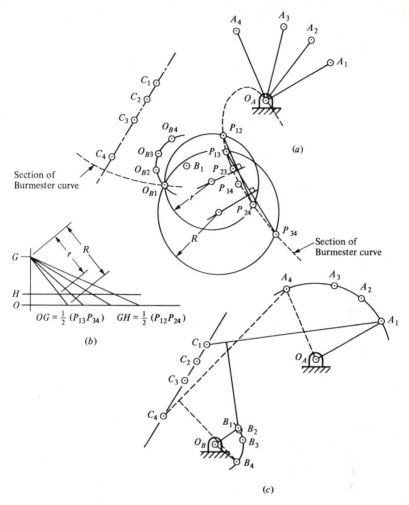

FIGURE 7.23
Burmester curve construction.

its four positions located on a straight line. For this example, pole lines $P_{12}P_{24}$ and $P_{13}P_{34}$ will be used to construct Burmester's curve (any of the above six sets of pole lines may be used). In order to locate the pole points, four positions of a coupler must be determined. Points C_1, C_2, C_3, and C_4 are located on the straight line with arbitrary distance $A_1C_1 = A_2C_2 = A_3C_3 = A_4C_4$. Using the four positions of a coupler line, the pole points may be located as shown.

Selecting the longest of the pole lines and laying off one-half the length on a vertical line locates point G in Fig. 7.23b. From point G, one-half the length of the shortest pole line is laid off to locate point H.

Horizontal lines are drawn from points G and H. A line from point G at any arbitrary angle will intersect the horizontal lines and indicate the radius of a circle drawn through the pole lines. Intersection of the two circles will locate a point on Burmester's curve and the curve is constructed.

Any point on Burmester's curve will serve as a fixed pivot point from a coupler point which will have its four positions located on a circle. Selecting point O_B as a fixed pivot point and inverting the mechanism (Fig. 7.23a) will locate the four inverted positions of O_B which will lie on the circle. Inversion is accomplished by making triangle $C_1A_1O_{B2}$ congruent to triangle $C_2A_2O_B$. Repeating the construction using positions 3 and 4 locates the inverted positions of O_B. Coupler point B is the center of the circle through inverted positions of O_B and the completed mechanism is shown in Fig. 7.23c.

7.12   OVERLAY

The overlay technique is well suited as a rapid technique for coordinating the angular displacement of an input crank and output rocker. Such mechanisms are best known as "function generators" since the crank and rocker angular displacements are related by some mathematical function. The technique is generally referred to as a "best fit" technique whose accuracy is determined by the patience of the designer.

The procedure is as follows. In Fig. 7.24a, the input crank is drawn in its specified positions to some convenient scale. Arcs b_1, b_2, b_3, b_4 of an arbitrary length are drawn from points A_1, A_2, A_3, and A_4. The point B must lie on these arcs in each of its four positions. A transparent overlay is made as shown in Fig. 7.24b with the proper angular displacements of the rocker and various arbitrary but known radii. The overlay is placed on the arcs which must contain the four positions of point B and moved until the intersection of arcs on the overlay align with the arcs which must contain the four positions of point B. When a satisfactory alignment is achieved, the length of the output rocker and location of its pivot point are noted (Fig. 7.24a). In the event that an alignment cannot be found, the arbitrary distance AB may be changed and/or the overlay may be turned upside down. In general, a good fit may be found in a very short period of time.

The overlay technique may be used for two to five positions with ease. When more than five positions are involved, the overlay technique becomes rather tedious and accuracy is generally sacrificed.

A generalized overlay based on displacements related by the seventh root of 2 has been developed by Professor Jeremy Hirschhorn at the University of New South Wales. The generalized overlay is generally suitable for as many as seven displacement positions. Use of the general-

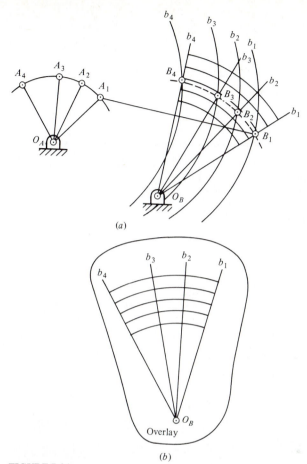

FIGURE 7.24
Overlay technique.

ized overlay requires that the input crank displacements be carefully selected.

7.13 THE VELOCITY POLE

The instant center for velocity between links 1 and 3 in Fig. 7.25 is located the distance p from center O_A. The velocity of point P_{13} considered as a point on link 1 must be the same as its velocity considered as a point on link 3. Therefore, with distance from O_A to O_B positive,

$$V_{P13} = p\omega_1 = (O_A O_B - p)\omega_3 \quad \text{crossed mechanism}$$

or $\quad p\omega_1 = (O_A O_B + p)\omega_3 \quad \text{open mechanism}$

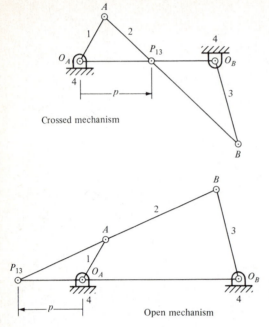

Crossed mechanism

Open mechanism

FIGURE 7.25
Velocity pole location.

Then, with the velocity ratio $VR = \omega_3/\omega_1$, the distance p becomes

$$p = \pm \frac{(VR)(O_A O_B)}{VR - 1}$$

where the plus sign refers to the crossed mechanism and the minus sign refers to the open mechanism.

In order to mechanize a function, it is necessary to perform some preliminary decisions and calculations before resorting to the graphical techniques.

Consider mechanization of the function $y = \ln x$ in the range of $x = 1$ to 10 with precision derivatives at $x = 1, 3, 7,$ and 10 (Fig. 7.26). The derivative of the function becomes $dy/dx = 1/x$ and a table is constructed.

Position	x	y	dy/dx
1	1	0.000	1.000
2	3	1.098	0.333
3	7	1.946	0.143
4	10	2.303	0.100

Allowing displacement of the input crank to be $90°$ and displacement of the output rocker to be $-90°$ (both assumed less than $120°$), the scale

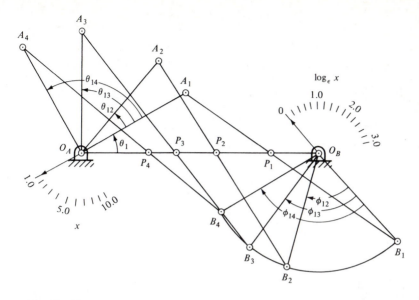

FIGURE 7.26
Function generation using the velocity pole.

factors may be computed. A crossed mechanism is more convenient to design and for this reason the negative sign is included in the output displacement.

$$k_\theta = \frac{\delta\theta}{\delta x} = \frac{90}{x_4 - x_1} = \frac{90}{10 - 1} = \frac{10°}{x}$$

$$k_\phi = \frac{\delta\phi}{\delta y} = \frac{-90}{y_4 - y_1} = \frac{-90}{2.303 - 0.000} = \frac{-39.08°}{y}$$

θ displacements

$x_{12} = 3 - 1 = 2$	$\theta_{12} = (10)2 = 20°$
$x_{13} = 7 - 1 = 6$	$\theta_{13} = (10)6 = 60°$
$x_{14} = 10 - 1 = 9$	$\theta_{14} = (10)9 = 90°$

ϕ displacements

$y_{12} = 1.098 - 0.000 = 1.098$	$\phi_{12} = (-39.08)1.098 = -42.91°$
$y_{13} = 1.946 - 0.000 = 1.946$	$\phi_{13} = (-39.08)1.946 = -76.05°$
$y_{14} = 2.303 - 0.000 = 2.303$	$\phi_{14} = (-39.08)2.303 = -90.00°$

The velocity ratio VR in terms of angular displacement rates is converted to include the differential of the function to be generated by use of the selected angular displacement scale factors.

$$\frac{\omega_3}{\omega_1} = \frac{d\phi/dt}{d\theta/dt} = \frac{d\phi}{d\theta} = \frac{k_\phi}{k_\theta}\frac{dy}{dx}$$

$$\mathrm{VR} = \frac{-39.08}{10.00}\frac{dy}{dx} = -3.908\frac{dy}{dx}$$

With the assumption that distance $O_A O_B = 10$, the distance p may be calculated.

$$p = \frac{(\mathrm{VR})(O_A O_B)}{\mathrm{VR} - 1}$$

$$p_1 = \frac{(-3.908)(1.000)(10)}{-3.908(1.000) - 1} = 7.96$$

$$p_2 = 5.65$$

$$p_3 = 3.58$$

$$p_4 = 2.81$$

Assuming that the distance $O_A A$ is equal to 5.0, the graphical procedure is initiated. The base link is constructed as in Fig. 7.26 (using a crossed mechanism). Initial position of the input crank is assumed, and the specified positions of point A are located using the selected input displacement angles.

Positions of the velocity pole are located according to the calculated p distances for a crossed mechanism. The distance AB is assumed and constructed through pole position p_1 from point A_1 to locate point B_1. The same AB distance is constructed through pole position p_4 from A_4 to locate point B_4. Points B_1 and B_4 must lie on a circle centered at O_B and must subtend the angle ϕ_{15} at O_B. If this is not the case, one or more of the assumptions must be altered to correct the situation.

When a fit is found (and it is generally a rapid solution), the intermediate positions are checked for their accuracy and modifications made as necessary. The designer must decide on how much accuracy is desired before concluding the problem.

7.14   THE ATLAS

Analysis of the Four Bar Linkage by Hrones and Nelson[*] is in reality an atlas of four-bar crank rocker mechanisms. The atlas is a large 11×17 page book which contains over 7000 coupler curves and dimensions of the mechanisms which will produce the curve.

---

[*] J. A. Hrones and G. L. Nelson, *Analysis of the Four Bar Linkage*, MIT Press and John Wiley & Sons, 1951.

The input crank is always the shortest link and of unity length, while the output rocker and coupler links vary from 1.5 to 4.0 in steps of 0.5. The fixed link varies from 1.5 to 6.5 in steps of 0.5. Coupler curves are provided for each of the 50 points on a grid attached to the coupler link. The coupler curves are constructed of dashed lines with the length of each dash representing 5° of rotation of the input crank. In using the atlas, a designer may thumb through the book to locate a curve of the desired shape. The length of the dashed lines and their spacing will give an indication of the coupler point velocity if the input crank angular velocity is constant. Consideration of the dashed lines as velocity vectors will allow some assessment of the direction and magnitude of the coupler point acceleration.

With the advent of computer programs, the atlas has lost some of its popularity and is out of print. A designer may now use an interactive program and generate coupler curves which may be modified until a desired curve is developed.

7.15 GRAPHICAL SYNTHESIS USING THE ACCELERATION POLE

For operation in steady state it is possible to synthesize a mechanism to match a specified coupler point location, velocity, and acceleration. Since the designer is generally aware of the input drive system or has the freedom to specify the drive system, the design of a mechanism to match a specified velocity and acceleration at a point in reality becomes a problem of designing a mechanism with known velocity and acceleration at two coupler points.

The technique requires use of the inflection circle which represents the instantaneous locus of all points on a coupler link which have zero normal acceleration. The Bresse circle represents the instantaneous locus of all points on the coupler link which have zero tangential acceleration. The cubic of stationary curvature represents the instantaneous locus of all points moving on a circular arc of constant radius. The inflection circle and the Bresse circle will intersect at two points representing the instantaneous centers for velocity and acceleration.

In all incidents, the location of the instant center for acceleration is never known. However, if one must design a mechanism to produce a specified coupler point location and acceleration during steady-state operation, selection of a constant input angular velocity and crank length of the driver provides a second coupler point with known velocity and acceleration.

With the acceleration of two coupler points known, it is possible to describe a quadrilateral such as *ABCD* in Fig. 7.27. From geometric considerations it is known that the four circles which circumscribe one

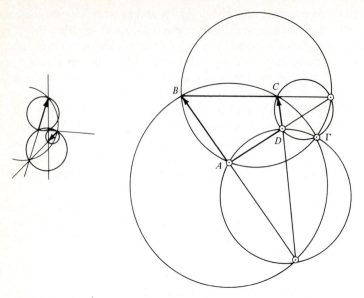

FIGURE 7.27
Four-circle construction to locate acceleration pole.

side and the apex of two adjacent sides of a quadrilateral will intersect at
a common point Γ. When the opposite sides of a quadrilateral are
acceleration vectors, the common intersection point is the instantaneous
center for acceleration. This is known as the four-circle construction
although only two circles are necessary to locate Γ.

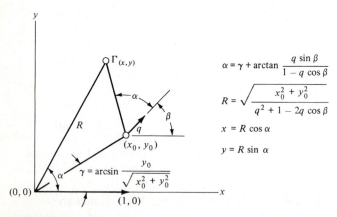

$$\alpha = \gamma + \arctan \frac{q \sin \beta}{1 - q \cos \beta}$$

$$R = \sqrt{\frac{x_0^2 + y_0^2}{q^2 + 1 - 2q \cos \beta}}$$

$$x = R \cos \alpha$$

$$y = R \sin \alpha$$

$$\gamma = \arcsin \frac{y_0}{\sqrt{x_0^2 + y_0^2}}$$

FIGURE 7.28
Analytic location of acceleration pole.

Analytically the x and y coordinates of the instantaneous center for acceleration may be found as shown in Fig. 7.28. To use the analytic equations it is necessary to establish the x axis of the coordinate system coincident with one of the known acceleration vectors and to make that vector of unit length. The other acceleration vector must be scaled accordingly to give the length q.

In Fig. 7.29 point A is the specified point with velocity $\mathbf{V}_A$ and acceleration $\mathbf{A}_A$. Point B is a coupler point associated with the input drive system and has velocity $\mathbf{V}_B$ and acceleration $\mathbf{A}_B$ which, in steady-state operation, are orthogonal. Location of point B as well as its velocity and acceleration vectors is an arbitrary design choice.

The instantaneous center for velocity, pole point P, is located at the intersection of orthogonals to the velocity vectors. The instantaneous center for acceleration, point Γ, may be located analytically or graphically in the following manner (Fig. 7.29).

1. Graphically determine the acceleration vector difference $\mathbf{A}_{A/B} = \mathbf{A}_A - \mathbf{A}_B$.

2. Determine the angle α between the acceleration vector difference $\mathbf{A}_{A/B}$ and the line AB. The positive sense is taken as the angle from the acceleration vector difference to the line AB.

3. Construct a line through A at the positive angle α with the acceleration vector $\mathbf{A}_A$. Construct another line through B at the positive angle α with the acceleration vector $\mathbf{A}_B$. Intersection of these two lines will be the instantaneous center for acceleration Γ.

4. The pole normal axis p_n is constructed through P at the negative angle α with the line $P\Gamma$. The pole tangent axis p_t is orthogonal with the pole normal axis at P.

5. The perpendicular bisector of $P\Gamma$ will intersect the pole normal axis at the center of the inflection circle and will intersect the pole tangent axis at the center of the Bresse circle.

Since the Bresse circle provides the locus of all points of the moving plane with zero tangential acceleration and the cubic of stationary curve provides the locus of all points of the moving plane with constant radius of curvature, their intersection will be a point moving on a circular arc with constant angular acceleration of the point's radius of curvature. Such a point is the moving hinge pin of an input crank.

In order to construct the cubic of stationary curvature (Fig. 7.29)

6. Construct a perpendicular to the line PB at point B. At the intersection with the pole normal axis construct a line parallel to the pole tangent axis. At the intersection with the pole tangent axis construct a line parallel with the pole normal axis.

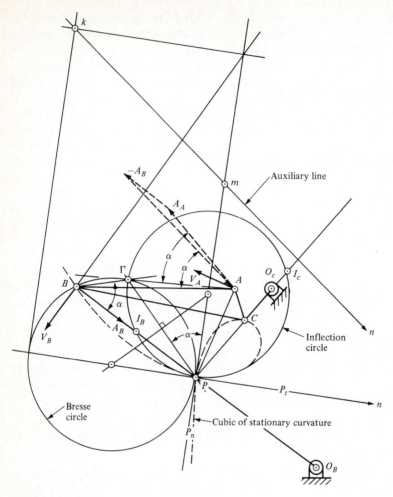

FIGURE 7.29
Synthesis using the acceleration pole.

7. The point k is found as a point on an auxiliary line used in construction of the cubic of stationary curvature. An arbitrary line through k will intersect the pole normal axis at m and the pole tangent axis at n thus describing the ratio m/n in the equation of the cubic of stationary curvature.

8. The cubic of stationary curvature is constructed by selecting arbitrary points along the auxiliary line through k.

9. The moving hinge pin of the output link is selected as a point on the cubic of stationary curvature (point C of Fig. 7.29).

10. The fixed hinge pin of the output link may be found through use of Bobillier's construction or by application of the Euler-Savary equation.

In the event that a suitable mechanism does not result, another location of C on the cubic of stationary curvature may be selected. If nothing appealing results, another slope of the auxiliary line through k can be chosen and another cubic of stationary curvature constructed. With a twofold infinity of mechanisms available, it is highly likely that an appealing mechanism can be found.

PROBLEMS

7.1. Lay out a crank rocker mechanism so that the output rocker will rotate through 60° and the minimum transmission angle is 40°.

7.2. Design a quick-return crank rocker mechanism with a transmission ratio of 1.125 and an output rocker displacement angle of 80°.

7.3. Design a quick-return crank rocker mechanism with a transmission ratio of 1.11 and an output rocker displacement of 70°. Initial crank angle is to be 17.5°.

7.4. For the two coupler link positions indicated by lines A_1B_1 and A_2B_2, locate the pole point. Using points C and D as moving hinge pins, design a four-bar mechanism that will move line AB into its two designated positions.

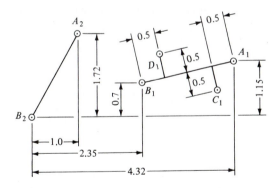

FIGURE P7.4
Two-position mechanism design.

7.5. The pole triangle for three positions of a coupler plane and the position of coupler point C when the coupler is in position 1 are given. Locate the position of C when the coupler is in positions 2 and 3. Use the rotation poles and displacement angles and check using the cardinal position of C.

$$P_{23}P_{13} = 2.00 \qquad P_{13}P_{12} = 2.50 \qquad P_{23}P_{12} = 2.25$$

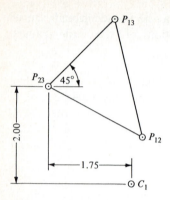

FIGURE 7.5
Rotation poles.

7.6. For the pole triangle given, locate the fixed hinge pin associated with coupler point C_2 and the moving hinge pin associated with fixed hinge pin O_D. Show that the three positions of C and D lie on circles. Coordinates are from the lower left corner of an $8\frac{1}{2} \times 11$ page.

$$P_{12} = (6.0, 6.0) \qquad P_{13} = (3.5, 5.0) \qquad P_{23} = (3.0, 7.0)$$
$$C_2 = (1.5, 5.0) \qquad O_D = (4.5, 4.0)$$

7.7. Locate a coupler point C_1 such that in the three positions of the mechanism indicated by the pole center triangle the three positions of C will lie on a straight line at an angle of 105° with the horizontal. Pole point coordinates are from the lower left corner of an $8\frac{1}{2} \times 11$ page.

$$P_{12} = (4.0, 5.0) \qquad P_{13} = (4.5, 5.5) \qquad P_3 = (3.5, 7.0)$$

7.8. Power input to the mechanism is to be by a fixed pneumatic cylinder operating in a straight line. Locate a suitable line of action for the cylinder.

$$O_A = (7.0, 5.0) \qquad O_A A = 1.0 \qquad AB = 3.5 \qquad O_B B = 2.0$$
$$O_B = (5.0, 7.5) \qquad \theta_1 = 120° \qquad \theta_2 = 150° \qquad \theta_3 = 180°$$

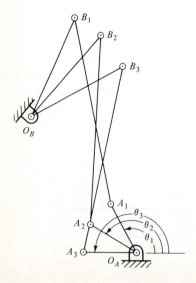

FIGURE 7.8
Coupler point with three positions on a straight line.

7.9. For the mechanism shown, select a suitable second fixed pivot O_C and locate the associated moving hinge pin C_1.

$$O_A = (3.0, 6.5) \qquad O_A A = 1.5 \qquad AB = 3.0$$
$$\theta_1 = 60° \qquad \theta_2 = 90° \qquad \theta_3 = 135° \qquad \theta_4 = 150°$$
$$\Gamma_1 = 85.5° \qquad \Gamma_2 = 89.0° \qquad \Gamma_3 = 101.0° \qquad \Gamma_4 = 104.0°$$

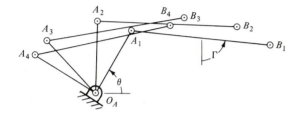

FIGURE P7.9
Four-position mechanism design.

7.10. For the four coupler positions indicated, construct a Burmester curve and locate an input crank and an output rocker.

$$CD = 3.0 \qquad C_1 = (1.55, 6.40) \qquad C_2 = (2.30, 7.70)$$
$$C_3 = (3.50, 8.80) \qquad C_4 = (4.50, 8.05) \qquad \Gamma_1 = 30°$$
$$\Gamma_2 = 0° \qquad \Gamma_3 = 300° \qquad \Gamma_4 = 240°$$

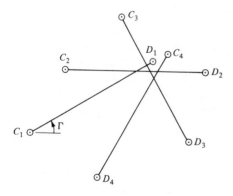

FIGURE P7.10
Burmester curve construction and use.

7.11. Lay out a mechanism to satisfy the following displacements:

Input crank	Output rocker
$\theta_{12} = 25°$	$\phi_{12} = -20°$
$\theta_{13} = 60°$	$\phi_{13} = -40°$

7.12. Lay out a mechanism to satisfy the following displacements:

Input crank	Output rocker
$\theta_{12} = 20°$	$\phi_{12} = 10°$
$\theta_{13} = 35°$	$\phi_{13} = 20°$
$\theta_{14} = 60°$	$\phi_{14} = 40°$

CHAPTER
8

ANALYTIC SYNTHESIS

8.1 INTRODUCTION

Analytic synthesis techniques have the advantage of very rapid solution of problems through the use of computer programs. Coupler curves can be displayed and modified as desired. The ability to animate the mechanism and observe its performance on a screen is a distinct advantage. At the same time the programs can plot velocity and acceleration curves in animation.

Many programs have been written and discussed in the literature. Unfortunately, the programs are generally directed at a specific type of synthesis or they are all-inclusive and quite long. Several programs are available commercially for use on main- or miniframe computers or on personal computers. Although the computer solutions are impressive, the user should be warned that their use can result in very time-consuming and expensive trial-and-error solutions.

Throughout this section the angles will be designated as θ_{ij} in which the subscript i refers to the link number and the subscript j refers to the position of the link. Thus θ_{23} would indicate the third position of link 2.

8.2 FREUDENSTEIN'S EQUATION

The works of Ferdinand Freudenstein in the early 1950s introduced the analytic techniques of mechanism synthesis. If the links of a mechanism are considered as directed line segments as in Fig. 8.1, a series of

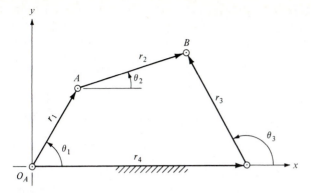

FIGURE 8.1
Notation for Freudenstein's equation.

equations may be written. In this analysis it is desirable to eliminate the coupler angle θ_2 since very little is known about that angle.

$$r_1 \cos \theta_1 + r_2 \cos \theta_2 = r_4 + r_3 \cos \theta_3$$

$$r_2 \cos \theta_2 = r_4 + r_3 \cos \theta_3 - r_1 \cos \theta_1 \qquad (8.1)$$

$$r_1 \sin \theta_1 + r_2 \sin \theta_2 = r_3 \sin \theta_3$$

$$r_2 \sin \theta_2 = r_3 \sin \theta_3 - r_1 \sin \theta_1 \qquad (8.2)$$

Squaring and adding Eqs. (8.1) and (8.2) and using trigonometric identities results in

$$r_2^2 = r_1^2 + r_3^2 + r_4^2 + 2r_4 r_3 \cos \theta_3 - 2r_4 r_1 \cos \theta_1 - 2r_1 r_3 \cos (\theta_1 - \theta_3)$$

Dividing by $2r_1 r_3$ and introducing the side ratios

$$R_1 = \frac{r_4^2 + r_3^2 + r_1^2 - r_2^2}{2r_1 r_3}$$

$$R_2 = \frac{r_4}{r_1}$$

$$R_3 = \frac{r_4}{r_3}$$

$$R_1 + R_2 \cos \theta_3 - R_3 \cos \theta_1 = \cos (\theta_1 - \theta_3) \qquad (8.3)$$

This is Freudenstein's equation which is useful in the synthesis of mechanisms of various types. More important, the technique used in deriving the equation forms the basis of many analytic synthesis techniques. If the equation is to be used for mechanism synthesis, it is interesting to note that only seven parameters are involved:

3 link ratios   2 angles   2 scale factors

Freudenstein's equation is particularly useful for synthesis of function generation mechanisms. It is necessary to make several decisions. The number of known or specified parameters plus the number of those decided must equal seven. For synthesis of a function generation mechanism with three precision positions, four decisions are allowed.

Example 8.1. The function $y = \cos x$ is to be mechanized with x ranging from 60 to 15° and with precision at $x = 50°$, 35°, and 20°.

Solution
In this case, y will vary from 0.966 to 0.500 with precision at $y = 0.643$, 0.819, and 0.940. With four arbitrary decisions allowed, the range of motion of the input crank is selected at $\Delta\theta_1 = 60°$, the range of motion of the output crank is selected as $\Delta\theta_3 = 100°$, initial position of the input crank is selected to be 45°, and initial position of the output crank is selected to be 30°.

The scale factors become

$$k\theta_1 = \frac{60}{60 - 15} = \frac{1.333°}{\text{unit } x}$$

$$k\theta_3 = \frac{100}{0.966 - 0.500} = \frac{214.6°}{\text{unit } y}$$

Precision positions of the cranks become

$$\theta_{11} = \theta_{1i} + k\theta_1(x_i - x_1) = 45 + 1.333(20 - 15) = 51.7°$$

$$\theta_{12} = 45 + 1.333(35 - 15) = 71.7°$$

$$\theta_{13} = 45 + 1.333(50 - 15) = 91.7°$$

$$\theta_{31} = \theta_{3i} + k\theta_3(y_i - y_1) = 30 + 214.6(0.966 - 0.940) = 35.6°$$

$$\theta_{32} = 30 + 214.6(0.966 - 0.819) = 61.5°$$

$$\theta_{33} = 30 + 214.6(0.966 - 0.643) = 99.3°$$

With these data, the Freudenstein equation may be written three times.

$$R_1 + 0.813R_2 - 0.620R_3 = 0.961$$

$$R_1 + 0.477R_2 - 0.314R_3 = 0.984$$

$$R_1 - 0.162R_2 + 0.030R_3 = 0.991$$

Simultaneous solution of these three equations provides

$$R_1 = \frac{r_4^2 + r_3^2 + r_1^2 - r_2^2}{2r_1 r_3} = 0.999$$

$$R_2 = \frac{r_4}{r_1} = 0.079$$

$$R_3 = \frac{r_4}{r_3} = 0.168$$

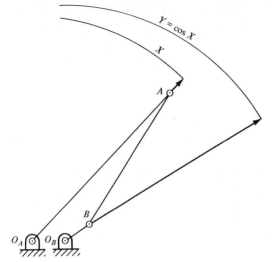

FIGURE 8.2
Solution for Example 8.1.

This procedure results in three equations with four unknowns. Solution is found by selecting one of the three unknowns to be a convenient size and solving for the remaining three values. Thus,

$$r_1 = 12.658 \qquad r_2 = 12.689$$
$$r_3 = 1.891 \qquad r_4 = 1.000$$

A mechanism with the above dimensions (Fig. 8.2) will perform as a function generator which mechanizes the function $y = \cos x$. However, it is not a very good mechanism since it is not a Grashof mechanism and has a poor initial transmission angle and branching is required. A more suitable mechanism may be realized by changing one or more of the initial decisions. It is wise to change only one decision at a time and note its influence rather than changing several of the decisions at once.

8.3 FREUDENSTEIN'S EQUATION WITH LEAST SQUARES

Synthesis of a function generator with three precision positions is relatively simple and fast using the Freudenstein equation. For those situations which require four or five precision positions, the Freudenstein equation becomes very cumbersome and time-consuming. In addition the accuracy of the resulting mechanism between the precision points may not be very satisfactory. For cases in which greater accuracy is desired between precision points, the Freudenstein equation may be combined with a least-squares technique. The least-squares application to kinematic synthesis probably started with Levitskii and Shakvasian in 1954.

The error in a function generation mechanism may be defined in terms of the desired output crank angle (θ_{3d}) and the obtained output crank angle (θ_{3o}).

$$\text{Error} = \frac{1}{n-1} \sum_{j=1}^{n} (\theta_{3dj} - \theta_{3oj})^2$$

where n is the number of positions of the mechanism and j is the position of the mechanism. The quantity $1/(n-1)$ may be omitted in the following since the equations will all be written as being equal to zero.

A is defined as (subscript j indicates position number)

$$A = \sum_{j=1}^{n} [R_1 + R_2 \cos \theta_{3j} - R_3 \cos \theta_{1j} - \cos (\theta_{1j} - \theta_{3j})]^2$$

If the error is to be minimized, the partial derivatives of A with respect to R_1, R_2, and R_3 may be equated to zero.

$$\frac{\partial A}{\partial R_1} = \sum_{j=1}^{n} [R_1 + R_2 \cos \theta_{3j} - R_3 \cos \theta_{1j} - \cos (\theta_{1j} - \theta_{3j})] = 0$$

or

$$R_1 + \sum_{j=1}^{n} R_2 \cos \theta_{3j} - \sum_{j=1}^{n} R_3 \cos \theta_{1j} - \sum_{j=1}^{n} \cos (\theta_{1j} - \theta_{3j}) = 0 \quad (8.4)$$

$$\frac{\partial A}{\partial R_2} = \sum_{j=1}^{n} [R_1 + R_2 \cos \theta_{3j} - R_3 \cos \theta_{1j} - \cos (\theta_{1j} - \theta_{3j})] \cos \theta_{3j} = 0$$

or

$$\sum_{j=1}^{n} R_1 \cos \theta_{3j} + \sum_{j=1}^{n} R_2 \cos^2 \theta_{3j} - \sum_{j=1}^{n} R_3 \cos \theta_{1j} \cos \theta_{3j}$$

$$- \sum_{j=1}^{n} \cos (\theta_{1j} - \theta_{3j}) \cos \theta_{3j} = 0 \quad (8.5)$$

$$\frac{\partial A}{\partial R_3} = \sum_{j=1}^{n} [R_1 + R_2 \cos \theta_{3j} - R_3 \cos \theta_{1j} - \cos (\theta_{1j} - \theta_{3j})] \cos \theta_{1j} = 0$$

or

$$\sum_{j=1}^{n} R_1 \cos \theta_{1j} + \sum_{j=1}^{n} R_2 \cos \theta_{3j} \cos \theta_{1j} - \sum_{j=1}^{n} R_3 \cos^2 \theta_{1j}$$

$$- \sum_{j=1}^{n} \cos (\theta_{1j} - \theta_{3j}) \cos \theta_{1j} = 0 \quad (8.6)$$

Equations (8.4) through (8.6) represent three equations in three unknown link ratios. Substitution of the known angular positions will allow solution for the link ratios. Selection of one of the link lengths will then give the link dimensions.

Example 8.2. The function $y = \cos x$ is to be mechanized with the angular displacements given in Table 8.1.

Solution
Using $k\theta_3 = 215°/\text{unit } y$ (which gives $\Delta\theta_3 = 100°$) and the initial positions of $\theta_1 = 15°$ and $\theta_3 = 30°$ Table 8.1 was constructed.

Equations (8.4), (8.5), and (8.6) become

$$R_1 + 2.660R_2 - 7.687R_3 = 7.806$$
$$2.660R_1 + 3.264R_2 - 2.905R_3 = 3.088$$
$$7.687R_1 + 2.905R_2 - 3.264R_3 = 6.300$$

Solution of the three equations results in the link ratios:

$$R_1 = 0.526 \qquad R_2 = -0.470 \qquad R_3 = -1.110$$

Selection of $r_1 = 1.0$ leads to

$$r_1 = 1.000 \qquad r_2 = 0.923$$
$$r_3 = 0.423 \qquad r_4 = -0.470$$

The negative sign on r_4 simply means that fixed center O_B is to the left of fixed center O_A. The mechanism resulting from the least-squares application to Freudenstein's equation (Fig. 8.3) is one which satisfies the Grashof criterion much better and has much improved transmission angles. Of course, the mechanism may be scaled up or down to suit the space available.

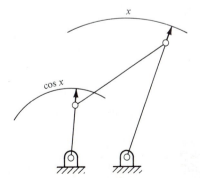

FIGURE 8.3
Solution for Example 8.2.

TABLE 8.1

Position	$x = \theta_1$	$y = \cos \theta_1$	θ_3	$\cos \theta_3$	$\cos^2 \theta_3$	$\cos^2 \theta_1$	$\cos (\theta_1 - \theta_3)$	$\cos \theta_1 \cos \theta_3$	$\cos (\theta_1 - \theta_3) \cos \theta_3$	$\cos (\theta_1 - \theta_3) \cos \theta_1$
1	60	0.500	130.0	-0.643	0.413	0.250	0.342	-0.322	-0.220	0.171
2	55	0.574	114.3	-0.412	0.170	0.329	0.511	-0.236	-0.211	0.293
3	50	0.643	99.4	-0.163	0.027	0.413	0.651	-0.105	-0.106	0.419
4	45	0.707	85.7	0.075	0.006	0.450	0.758	0.053	0.057	0.536
5	40	0.766	73.0	0.292	0.085	0.587	0.839	0.224	0.245	0.643
6	35	0.819	61.6	0.476	0.227	0.671	0.894	0.390	0.426	0.732
7	30	0.866	51.5	0.623	0.388	0.750	0.930	0.540	0.579	0.805
8	25	0.906	42.9	0.733	0.537	0.828	0.952	0.664	0.698	0.863
9	20	0.940	35.6	0.813	0.661	0.884	0.963	0.764	0.783	0.905
10	15	0.966	30.0	0.866	0.750	0.933	0.966	0.933	0.837	0.933
Total		7.687		2.660	3.264	6.088	7.806	2.905	3.088	6.300

8.4 FREUDENSTEIN'S EQUATION FOR ANGULAR VELOCITIES AND ACCELERATIONS

By differentiation of the Freudenstein equation, it is possible to achieve a third-order approximation at a particular phase of the mechanism. By continued differentiation, fourth- and fifth-order approximations are possible, but the procedure is rather tedious.

Defining θ_3' as $d\theta_3/d\theta_1$ and remembering that $d\theta_3 = k\theta_3 \, dy$ and $d\theta_1 = k\theta_1 \, dx$:

$$\theta_3' = \frac{d\theta_3}{d\theta_1} = \frac{d\theta_3}{dy} \frac{dy}{dx} \frac{dx}{d\theta_1} = \frac{k\theta_3 \, dy}{dy} \frac{dy}{dx} \frac{dx}{k\theta_1 \, dx} = \frac{k\theta_3}{k\theta_1} y'$$

$$\theta_3'' = \frac{d\theta_3'}{d\theta_1} = \frac{d\theta_3'}{dy'} \frac{dy'}{dx} \frac{dx}{d\theta_1} = \frac{k\theta_3}{k\theta_1^2} y''$$

With these preliminaries Eq. (8.3) may be differentiated twice to provide three equations in R_1, R_2, and R_3 which may be solved simultaneously and the link dimensions determined.

$$R_1 + R_2 \cos \theta_3 - R_3 \cos \theta_1 = \cos (\theta_1 - \theta_3) \tag{8.7}$$

The first derivative becomes

$$-R_2 \theta_3' \sin \theta_3 + R_3 \sin \theta_1 = -(1 - \theta_3') \sin (\theta_1 - \theta_3) \tag{8.8}$$

The second derivative becomes

$$-R_2[(\theta_3')^2 \cos \theta_3 + \theta_3'' \sin \theta_3] + R_3 \cos \theta_1$$
$$= \theta_3'' \sin (\theta_1 - \theta_3) - (1 - \theta_3')^2 \cos (\theta_1 - \theta_3) \tag{8.9}$$

For a third-order approximation, the scale factors $k\theta_1$ and $k\theta_3$ as well as the mechanism position where the approximation is to take place (θ_{1p} and θ_{3p}) are known. After calculation of θ_3' and θ_3'', Eqs. (8.7) through (8.9) will be three equations with three unknown link ratios, and the link dimensions may be determined.

The resulting mechanism must be scrutinized for transmission angles and excessive link ratios as well as for compliance with the Grashof criterion before being accepted. If the mechanism is not satisfactory for any reason, it is necessary to repeat the process using a different decision at the outset.

8.5 COMPONENTS

The component technique is an algebraic approach in which the configuration diagram, the velocity polygon, and the acceleration polygon provide six equations with thirteen unknowns. The unknowns include the fixed link length and the x and y components of the other three link lengths, three angular velocities, and three angular accelerations. If seven parameters can be specified, then the six equations may be solved for the remaining six unknowns.

A mechanism configuration diagram and the associated velocity and acceleration polygons are shown in Fig. 8.4.

From the configuration diagram

$$r_{1x} + r_{2x} = r_4 + r_{3x} \tag{8.10}$$

$$r_{1y} + r_{2y} = r_{3y} \tag{8.11}$$

From the velocity polygon in which $V_A + V_{B/A} = V_B$,

$$V_A = r_1 \omega_1 \qquad V_{B/A} = r_2 \omega_2 \qquad V_B = r_3 \omega_3$$

$$r_{1x} \omega_1 + r_{2x} \omega_2 = r_{3x} \omega_3 \tag{8.12}$$

$$r_{1y} \omega_1 + r_{2y} \omega_2 = r_{3y} \omega_3 \tag{8.13}$$

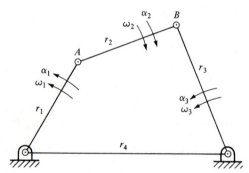

Configuration diagram

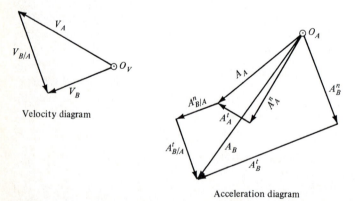

Velocity diagram

Acceleration diagram

FIGURE 8.4
Notation for method of components.

From the acceleration polygon in which $A_B = A_A + A_{B/A}$,

$$A_A^n = r_1\omega_1^2 \qquad A_A^t = r_1\alpha_1$$

$$A_B^n = r_3\omega_3^2 \qquad A_B^t = r_3\alpha_3$$

$$A_{B/A}^n = r_2\omega_2^2 \qquad A_{B/A}^t = r_2\alpha_2$$

$$r_{1x}\omega_1^2 + r_{1y}\alpha_1 + r_{2x}\omega_2^2 + r_{2y}\alpha_2 = r_{3x}\omega_3^2 + r_{3y}\alpha_3 \qquad (8.14)$$

$$r_{1y}\omega_1^2 - r_{1x}\alpha_1 + r_{2y}\omega_2^2 - r_{2x}\alpha_2 = r_{3y}\omega_3^2 - r_{3x}\alpha_3 \qquad (8.15)$$

Equations (8.10) through (8.15) may be used in various ways for synthesis problems in which seven of the thirteen parameters are known.

The technique is particularly handy and very fast for synthesis of a mechanism with specified angular velocity and angular acceleration of each link. In this case it is only necessary to specify the length of the fixed link which allows writing of six equations with six unknowns. The six equations may be solved by any convenient means.

Example 8.3. A mechanism is to be designed to provide

$$r_4 = 10 \qquad \omega_1 = 20 \qquad \omega_2 = 10 \qquad \omega_3 = 15$$

$$\alpha_1 = 0 \qquad \alpha_2 = 5 \qquad \alpha_3 = 8$$

Solution
The six equations become

$$r_{1x} + 0r_{1y} + r_{2x} + 0r_{2y} - r_{3x} - 0r_{3y} = 10$$

$$0r_{1x} + r_{1y} + 0r_{2x} + r_{2y} - 0r_{3x} - r_{3y} = 0$$

$$20r_{1x} + 0r_{1y} + 10r_{2x} + 0r_{2y} - 15r_{3x} - 0r_{3y} = 0$$

$$0r_{1x} + 20r_{1y} + 0r_{2x} + 10r_{2y} - 0r_{3x} - 15r_{3y} = 0$$

$$20^2r_{1x} + 0r_{1y} + 10^2r_{2x} + 5r_{2y} - 15^2r_{3x} - 8r_{3y} = 0$$

$$-0r_{1x} + 20^2r_{1y} - 5r_{2x} + 10^2r_{2y} + 8r_{3x} - 15^2r_{3y} = 0$$

Simultaneous solution of these six equations provides

$$r_4 = 10.000 \qquad r_4 = 10.000$$

$$r_{1x} = -5.720 \qquad r_1 = 5.187$$

$$r_{1y} = 1.058$$

$$r_{2x} = 24.280 \qquad r_2 = 24.303$$

$$r_{2y} = 1.058$$

$$r_{3x} = 8.560 \qquad r_3 = 8.818$$

$$r_{3y} = 2.117$$

The resulting mechanism (Fig. 8.5) meets the Grashof criterion but has a very bad transmission angle and probably branching. However, the

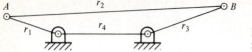

FIGURE 8.5
Solution for Example 8.3.

mechanism does meet the angular velocity and angular acceleration requirements.

In using the component technique, the general procedure is to eliminate unknowns from the six equations until a set of wieldy equations results. This often involves much algebraic manipulation, and for this reason use of the technique is limited.

In eliminating unknowns it is often convenient to proceed as follows:

From Eqs. (8.10) and (8.12)

$$r_{1x} = \frac{r_{3x}(\omega_2 - \omega_3) + r_4\omega_2}{\omega_2 - \omega_1} \tag{8.16}$$

$$r_{2x} = \frac{r_{2x}(\omega_2 - \omega_3) + r_4\omega_1}{\omega_1 - \omega_2} \tag{8.17}$$

From Eqs. (8.11) and (8.13)

$$r_{1y} = \frac{r_{3y}(\omega_2 - \omega_3)}{\omega_1 - \omega_2} \tag{8.18}$$

$$r_{2y} = \frac{r_{3y}(\omega_1 - \omega_3)}{\omega_1 - \omega_2} \tag{8.19}$$

Generally the angular velocity and acceleration of the coupler link (r_2) are not known. Elimination of α_2 between Eqs. (8.14) and (8.15) and substitution of Eqs. (8.16) through (8.19) results in an equation of the form

$$A\omega_2^2 + B\omega_2 + C = 0$$

in which

$$A = 2r_{3x}r_{3y}(\omega_1 - \omega_3)^2 + 2r_4r_{3y}\omega_1(\omega_1 - \omega_3)$$

$$B = r_{3y}r_4\omega_3\omega_1^2 + \alpha_1[r_{3y}^2(\omega_3 - \omega_1) + r_{3x}^2(\omega_1 - \omega_3)$$

$$- r_4r_{3x}(\omega_1 - \omega_3) - r_4^2\omega_1]$$

$$+ \omega_3^2[2r_{3x}r_{3y}(\omega_1 - \omega_3) + r_4\omega_1]$$

$$+ \alpha_3r_{3y}^2[(1 + r_{3y}^2)(\omega_1 - \omega_3) + r_4r_{3x}\omega_1]$$

$$C = -\omega_1 \omega_3^2 [2r_{3y} r_{3y}(\omega_1 - \omega_3) + r_4 \omega_1] + r_4 r_{3y} \omega_1^3 \omega_3$$
$$+ \alpha_3 \omega_1 [(r_{3x}^2 - r_{3y}^2)(\omega_1 - \omega_3) + r_4 r_{3x} \omega_1]$$
$$+ \alpha_1 [r_{3y}^2 (\omega_1 \omega_3 + \omega_3^2) - r_{3x}^2 \omega_3 (\omega_1 - \omega_3)$$
$$+ r_4 r_{3y} \omega_1 \omega_3]$$

In general, this equation may be solved for ω_2 directly. There will be two solutions, one of which will lead to infinitely long links and the other which may be useful. In the event that an output crank angle is one of the specifications, it will be necessary to substitute for r_{3y} in terms of r_{3x} and then to select a value for r_{3x} and solve for ω_2. Once ω_2 is known, the other unknowns are readily available. In the event that the resulting mechanism is unsatisfactory, it will be necessary to select a different value for r_{3x} and repeat the calculations. A table of results will provide guidance in selecting r_{3x}.

8.6 LOOP EQUATIONS

By considering the links of a mechanism as vectors, loop equations may be written. In Fig. 8.6, each link is designated as a vector. Starting at O_A and progressing to A, B, O_B, and back to O_A completes a loop. The loop equation becomes

$$r_1 e^{i\theta_1} + r_2 e^{i\theta_2} - r_3 e^{i\theta_3} - r_4 = 0 \qquad (8.20)$$

Application of Eq. (8.20) to the synthesis of a function generator with three positions requires writing the equation 3 times representing the three positions of the mechanism.

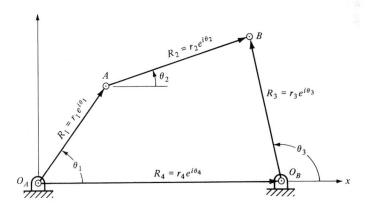

FIGURE 8.6
Notation for loop equations.

First position: $r_1 e^{i\theta_{11}} + r_2 e^{i\theta_{21}} - r_3 e^{i\theta_{31}} = r_4$

Second position: $r_1 e^{i\theta_{12}} + r_2 e^{i\theta_{22}} - r_3 e^{i\theta_{32}} = r_4$

Third position: $r_1 e^{i\theta_{13}} + r_2 e^{i\theta_{23}} - r_3 e^{i\theta_{33}} = r_4$

The initial position of the mechanism may be specified by selecting θ_{11} and θ_{31} which leaves only the length of the three links and the angular positions of the coupler link as unknowns. By judiciously selecting the angular positions of the coupler link, the only remaining unknowns are the lengths of the input link, the coupler link, and the output link.

Example 8.4. The function $y = \cos x$ is to be mechanized as in Sec. 8.2.

Solution
In this case

$$\theta_{11} = 21.7 \qquad \theta_{12} = 41.7 \qquad \theta_{13} = 61.7$$
$$\theta_{31} = 35.6 \qquad \theta_{32} = 61.5 \qquad \theta_{33} = 99.3$$

Selected values are

$$r_4 = 1.0 \qquad \theta_{21} = 60.0 \qquad \theta_{22} = 55.0 \qquad \theta_{23} = 50.0$$

By assigning the value of 1.0 to r_4, the three loop equations become

$$r_1 e^{i\theta_{11}} + r_2 e^{i\theta_{21}} + r_3 e^{i\theta_{31}} = r_4 = 1.0$$
$$r_1 e^{i\theta_{12}} + r_2 e^{i\theta_{22}} + r_3 e^{i\theta_{32}} = r_4 = 1.0$$
$$r_1 e^{i\theta_{13}} + r_2 e^{i\theta_{23}} + r_3 e^{i\theta_{33}} = r_4 = 1.0$$

This system of equations may be solved using Cramer's rule.

$$D = \begin{vmatrix} e^{i\theta_{11}} & e^{i\theta_{21}} & e^{i\theta_{31}} \\ e^{i\theta_{12}} & e^{i\theta_{22}} & e^{i\theta_{32}} \\ e^{i\theta_{13}} & e^{i\theta_{23}} & e^{i\theta_{33}} \end{vmatrix} = a + bi$$

$$r_1 = \frac{\begin{vmatrix} 1 & e^{i\theta_{21}} & e^{i\theta_{31}} \\ 1 & e^{i\theta_{22}} & e^{i\theta_{32}} \\ 1 & e^{i\theta_{23}} & e^{i\theta_{33}} \end{vmatrix}}{D} = \frac{c + di}{a + bi} \frac{a - bi}{a - bi}$$

$$r_2 = \frac{\begin{vmatrix} e^{i\theta_{11}} & 1 & e^{i\theta_{31}} \\ e^{i\theta_{12}} & 1 & e^{i\theta_{32}} \\ e^{i\theta_{13}} & 1 & e^{i\theta_{33}} \end{vmatrix}}{D} = \frac{f + gi}{a + bi} \frac{a - bi}{a - bi}$$

$$r_3 = \frac{\begin{vmatrix} e^{i\theta_{11}} & e^{i\theta_{21}} & 1 \\ e^{i\theta_{12}} & e^{i\theta_{22}} & 1 \\ e^{i\theta_{13}} & e^{i\theta_{23}} & 1 \end{vmatrix}}{D} = \frac{h + ji}{a + bi} \frac{a - bi}{a - bi}$$

where

$$i = \sqrt{-1}$$

$$a = \cos\left(\theta_{13} + \theta_{22} + \theta_{31}\right) + \cos\left(\theta_{11} + \theta_{23} + \theta_{32}\right) + \cos\left(\theta_{12} + \theta_{21} + \theta_{33}\right)$$
$$- \cos\left(\theta_{11} + \theta_{22} + \theta_{33}\right) - \cos\left(\theta_{13} + \theta_{21} + \theta_{32}\right) - \cos\left(\theta_{12} + \theta_{23} + \theta_{31}\right)$$

$$b = \sin\left(\theta_{13} + \theta_{22} + \theta_{31}\right) + \sin\left(\theta_{11} + \theta_{23} + \theta_{32}\right) + \sin\left(\theta_{12} + \theta_{21} + \theta_{33}\right)$$
$$- \sin\left(\theta_{11} + \theta_{22} + \theta_{33}\right) - \sin\left(\theta_{13} + \theta_{21} + \theta_{32}\right) - \sin\left(\theta_{12} + \theta_{23} + \theta_{31}\right)$$

$$c = \cos\left(\theta_{31} + \theta_{22}\right) + \cos\left(\theta_{23} + \theta_{32}\right) + \cos\left(\theta_{21} + \theta_{33}\right)$$
$$- \cos\left(\theta_{22} + \theta_{33}\right) - \cos\left(\theta_{23} + \theta_{31}\right) - \cos\left(\theta_{21} + \theta_{32}\right)$$

$$d = \sin\left(\theta_{31} + \theta_{22}\right) + \sin\left(\theta_{23} + \theta_{32}\right) + \sin\left(\theta_{21} + \theta_{33}\right)$$
$$- \sin\left(\theta_{22} + \theta_{33}\right) - \sin\left(\theta_{23} + \theta_{31}\right) - \sin\left(\theta_{21} + \theta_{33}\right)$$

$$f = \cos\left(\theta_{13} + \theta_{31}\right) + \cos\left(\theta_{11} + \theta_{23}\right) + \cos\left(\theta_{12} + \theta_{21}\right)$$
$$- \cos\left(\theta_{11} + \theta_{22}\right) - \cos\left(\theta_{12} + \theta_{23}\right) - \cos\left(\theta_{13} + \theta_{21}\right)$$

$$g = \sin\left(\theta_{13} + \theta_{31}\right) + \sin\left(\theta_{11} + \theta_{23}\right) + \sin\left(\theta_{12} + \theta_{21}\right)$$
$$- \sin\left(\theta_{11} + \theta_{22}\right) - \sin\left(\theta_{12} + \theta_{23}\right) - \sin\left(\theta_{13} + \theta_{21}\right)$$

$$h = \cos\left(\theta_{13} + \theta_{22}\right) + \cos\left(\theta_{11} + \theta_{23}\right) + \cos\left(\theta_{12} + \theta_{21}\right)$$
$$- \cos\left(\theta_{11} + \theta_{22}\right) - \cos\left(\theta_{12} + \theta_{23}\right) - \cos\left(\theta_{13} + \theta_{21}\right)$$

$$j = \sin\left(\theta_{13} + \theta_{22}\right) + \sin\left(\theta_{11} + \theta_{23}\right) + \sin\left(\theta_{12} + \theta_{21}\right)$$
$$- \sin\left(\theta_{11} + \theta_{22}\right) - \sin\left(\theta_{12} + \theta_{23}\right) - \sin\left(\theta_{13} + \theta_{21}\right)$$

Substituting the angle values and evaluating gives

$$r_1 = 0.134 - 0.232i \qquad r_1 = 0.268 \qquad r_4 = 1.0$$
$$r_2 = 0.118 - 0.606i \qquad r_2 = 0.617$$
$$r_3 = 0.071 + 0.082i \qquad r_3 = 0.108$$

This mechanism may be scaled up or down without changing the angular displacements of the links. If the mechanism is unsatisfactory for any reason, selection of different values for θ_{21}, θ_{22}, and θ_{23} will provide a different and possibly more satisfactory solution.

8.7 LOOP EQUATIONS FOR TWO POSITIONS FROM THE INITIAL POSITION WITH SPECIFIED COUPLER ANGLES

The problem of synthesizing a mechanism to assume three specified positions with designated coupler angles at those positions is encountered in many instances. The case of moving a manufactured part or carton from one position to another in the manufacturing process while avoiding obstacles in the path is frequently encountered. In the example following, the problem is to open a rather large valve plate and at the same time to direct vapor flow through the valve opening. Three is the maximum number of positions for which a mechanism may be synthesized with a linear solution of the equations. More than three positions (up to five) are solved with nonlinear techniques.

In Fig. 8.7 the mechanism is shown in two positions. Each link is indicated in its original position using capital letters. Thus R_1, R_2, R_3, R_4, R_5, and R_6 are the links of the mechanism, each of which is treated as a vector. For this reason the angular displacements are all measured from the initial position of the mechanism. Thus θ_{12} represents the displacement of link 1 from its initial position to position 2. Displacement of point C from its original position to another position is also treated as a vector.

In considering the jth position of the mechanism and links 1 and 5, a loop equation may be written.

$$R_1 e^{i\theta_{1j}} + R_5 e^{i\theta_{5j}} - \delta_j - R_5 - R_1 = 0$$

or

$$R_1(e^{i\theta_{1j}} - 1) + R_5(e^{i\theta_{5j}} - 1) = \delta_j \tag{8.21}$$

Considering links 3 and 6, another loop equation may be written.

$$R_3 e^{i\theta_{3j}} + R_6 e^{i\theta_{6j}} - \delta_j - R_6 - R_3 = 0$$

or

$$R_3(e^{i\theta_{3j}} - 1) + R_6(e^{i\theta_{6j}} - 1) = \delta_j \tag{8.22}$$

Note: $\theta_{6j} = \theta_{5j}$ since links 5 and 6 are two sides of the same rigid link. Links R_2, R_5, and R_6 comprise another loop.

$$R_2 + R_6 = R_5 \tag{8.23}$$

The last loop is made up of links R_1, R_2, R_3, and R_4.

$$R_4 = R_1 + R_2 - R_3 \tag{8.24}$$

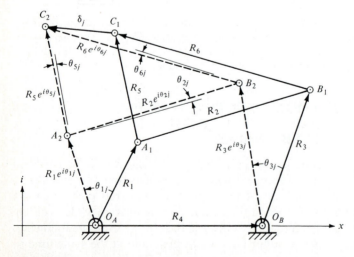

FIGURE 8.7
Loop equations from initial position.

Equations (8.21) through (8.24) may be used to synthesize a mechanism for specified positions of the mechanism with specified angular positions of the coupler link. With three positions, the solution is a linear solution. With more than three positions, the solution is a nonlinear solution. Fortunately, three specified positions are generally sufficient for most problems.

Example 8.5. Figure 8.8 shows a rather large valve plate which is to be opened and closed using a four-bar mechanism as shown. In order to operate properly the input crank is required to rotate through a total displacement of 150°. With displacements measured from the open position the following angular displacements are to be used in Eqs. (8.21) through (8.24). Note that θ_{12}, θ_{13}, θ_{32}, and θ_{33} are selected values which are varied to provide a satisfactory solution.

$$\theta_{12} = 30° \qquad \theta_{32} = 20° \qquad \theta_{52} = 9°$$
$$\theta_{13} = 45° \qquad \theta_{33} = 27° \qquad \theta_{53} = 36°$$
$$\delta_2 = 4'' \qquad \alpha_2 = 83°$$
$$\delta_3 = 10'' \qquad \alpha_3 = 100°$$

Equation (8.21) may be written twice:

$$R_1(e^{i30} - 1) + R_5(e^9 - 1) = 4e^{i83}$$
$$R_1(e^{i45} - 1) + R_5(e^{i36} - 1) = 10e^{i100} \tag{8.25}$$

Equations (8.25) may be solved for R_1 and R_5 using Cramer's rule or the computer program LOOPS.

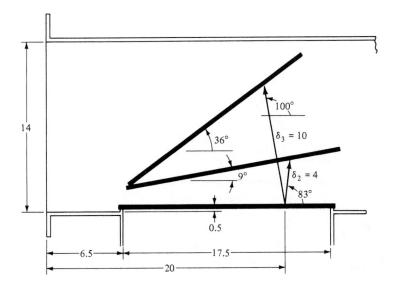

FIGURE 8.8
Valve plate problem.

Equations (8.22) may be written twice remembering that $\theta_{6j} = \theta_{5j}$:

$$R_3(e^{i20} - 1) + R_6(e^{i9} - 1) = 4e^{i83}$$
$$R_3(e^{i27} - 1) + R_6(e^{i36} - 1) = 10e^{i100} \tag{8.26}$$

Equations (8.26) may be solved for R_3 and R_6 after which Eqs. (8.23) and (8.24) may be used to solve for R_2 and R_4.

Output from the program LOOPS (App. A) is as follows:

| | LINK COMPONENTS | | |
LINK NO.	REAL	IMAGINARY	TOTAL LENGTH
1	3.8620	−2.3788	4.5358
2	−0.9480	−0.0341	0.9486
3	5.6155	−3.0047	6.3689
4	−2.7015	0.5919	2.7655
5	11.0261	0.3050	11.0304
6	11.9741	0.3391	11.9789

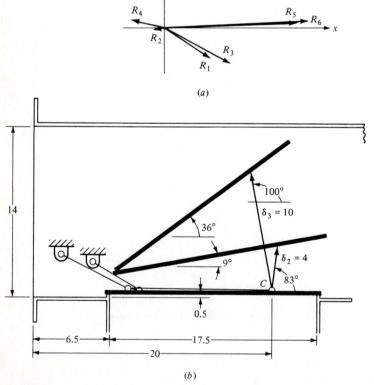

(a)

(b)

FIGURE 8.9
Valve plate problem solution.

The calculated links are plotted as shown in Fig. 8.9a and are shown assembled in their proper order in Fig. 8.9b.

8.8 GROUND LOOP EQUATIONS

Figure 8.10 shows a four-bar mechanism in relation to ground. Positions of the coupler point C are designated by vector $\mathbf{D} = (de^{i\delta_j})$. Ground pivots of the mechanism are located by vectors $\mathbf{A} = ae^{i\alpha}$ and $\mathbf{B} = be^{i\beta}$. Links of the mechanism are designated by vectors $\mathbf{R}_i$ with $i = 1$ to 6. Angular positions of the links are designated by angles θ_{ij} where j refers to the position number.

Starting at the coordinate center O a loop equation may be written for the path OO_AAC.

$$ae^{i\alpha} + r_1 e^{i\theta_{1j}} + r_5 e^{i\theta_{5j}} = d_j e^{i\delta_j} \tag{8.27}$$

Starting at the coordinate center O a loop equation may be written for the path OO_BBC.

$$be^{i\beta} + r_3 e^{i\theta_{3j}} + r_6 e^{i\theta_{6j}} = d_j e^{i\delta_j} \tag{8.28}$$

Equations (8.27) and (8.28) may be separated into their real and imaginary components.

$$a \cos \alpha + r_1 \cos \theta_{1j} + r_5 \cos \theta_{5j} = dj \cos \delta_j \tag{8.29}$$

$$a \sin \alpha + r_1 \sin \theta_{1j} + r_5 \sin \theta_{5j} = dj \sin \delta_j \tag{8.30}$$

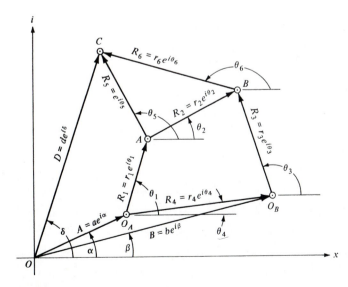

FIGURE 8.10
Notation for ground loop equations.

$$b \cos \beta + r_3 \cos \theta_{3j} + r_6 \cos \theta_{6j} = dj \cos \delta_j \qquad (8.31)$$

$$b \sin \beta + r_3 \sin \theta_{3j} + r_6 \sin \theta_{6j} = dj \sin \delta_j \qquad (8.32)$$

In these equations the link lengths r_1 through r_6 are unknown. Distances a and b may be selected along with angles α and β. The distance d and angles δ are specified by the problem to be solved. This leaves the angles θ_3, θ_5, and θ_6 to be completely unknown, although angles θ_5 and θ_6 are related by the angular displacement of the coupler link.

Solving Eq. (8.29) for $r_5 \cos \theta_{5j}$ and squaring, then solving Eq. (8.30) for $r_5 \sin \theta_{5j}$ and squaring, and adding will eliminate the angle θ_{5j} resulting in

$$a[2d_j \cos (\delta_j - \alpha)] + r_1[2d_j \cos (\theta_{1j} - \delta_j)] + (r_5^2 - a^2 - r_1^2)$$
$$= d_j^2 + ar_1[2 \cos (\theta_{1j} - \alpha)] \quad (8.33)$$

Solving Eq. (8.31) for $r_3 \cos \theta_{3j}$ and squaring, then solving Eq. (8.32) for $r_3 \sin \theta_{3j}$ and squaring, and adding will eliminate the angle θ_{3j} resulting in

$$b[2d_j \cos (\delta_j - \beta)] + r_6[2d_j \cos (\theta_{6j} - \delta_j)] + (r_3^2 - b^2 - r_6^2)$$
$$= d_j^2 + br_6[2 \cos (\theta_{6j} - \beta)] \quad (8.34)$$

Solution of Eqs. (8.33) and (8.34) is simplified by substitution of

$$k_1 = a \qquad\qquad\qquad k_5 = b$$
$$k_2 = r_1 \qquad\qquad\qquad k_6 r_6$$
$$k_3 = r_5^2 - a^2 - r_1^2 \qquad k_7 = r_3^2 - b^2 - r_6^2$$
$$k_4 = ar_1 = k_1 k_2 = \mu \qquad k_8 = br_6 = k_5 k_6 = \tau$$

With these substitutions Eqs. (8.33) and (8.34) become

$$k_1[2d_j \cos (\delta_j - \alpha)] + k_2[2d_j \cos (\theta_{1j} - \delta_j)] + k_3$$
$$= d_j^2 + k_4[2 \cos (\theta_{1j} - \alpha)] \quad (8.35)$$
$$k_5[2d_j \cos (\delta_j - \beta)] + r_6[2d_j \cos (\theta_{6j} - \delta_j)] + k_7$$
$$= d_j^2 + k_8[2 \cos (\theta_{6j} - \beta)] \quad (8.36)$$

Solution of Eqs. (8.35) and (8.36) is complicated by the fact that $k_4 = k_1 k_2$ and $k_8 = k_5 k_6$. The procedure is to introduce pseudo imaginary numbers of the form $e + \mu f$ and $g + \tau h$. Thus

$$k_1 = e_1 + \mu f_1 \qquad k_5 = g_5 + \tau h_5$$
$$k_2 = e_2 + \mu f_2 \qquad k_6 = g_6 + \tau h_6$$

$$k_3 = e_3 + \mu f_3 \qquad k_7 = g_7 + \tau h_7$$

$$k_4 = \mu \qquad\qquad k_8 = \tau$$

Making this substitution into Eqs. (8.35) and (8.36) and separating the equations into those parts containing μ and τ and those parts not containing μ and τ provides four additional equations.

$$e_1[2d_j \cos(\delta_j - \alpha)] + e_2[2d_j \cos(\theta_{1j} - \delta_j)] + e_3 = d_j^2 \qquad (8.37)$$

$$f_1[2d_j \cos(\delta_j - \alpha)] + f_2[2d_j \cos(\theta_{1j} - \delta_j)] + f_3 = 2\cos(\theta_{1j} - \alpha) \qquad (8.38)$$

$$g_5[2d_j \cos(\delta_j - \beta)] + g_6[2d_j \cos(\theta_{6j} - \delta_j)] + g_7 = d_j^2 \qquad (8.39)$$

$$h_5[2d_j \cos(\delta_j - \beta)] + h_6[2d_j \cos(\theta_{6j} - \delta_j)] + h_7 = 2\cos(\theta_{6j} - \beta) \qquad (8.40)$$

With three positions of the coupler point C specified along with three corresponding positions of θ_1, selection of a value for α will allow writing Eq. (8.37) 3 times and solving for e_1, e_2, and e_3. Equation (8.38) may be written 3 times and solved for f_1, f_2, and f_3. Then k_1, k_2, and k_3 may be evaluated. However, the compatibility condition must be used to solve for μ and since $\mu = k_1 k_2$, a quadratic equation results and two values of μ are determined leading to two separate solutions for a, r_1, and r_5. At the same time, using Eq. (8.29), the three values of θ_{51}, θ_{52}, and θ_{53} may be found.

With the same three positions of coupler point C known and selection of β, Eq. (8.39) may be written 3 times and solved for g_5, g_6, and g_7. In order to write the equations the value of θ_{6j} must be determined. The value of θ_{61} is selected. Then

$$\theta_{62} = \theta_{61} + (\theta_{52} - \theta_{51})$$

$$\theta_{63} = \theta_{61} + (\theta_{53} - \theta_{51})$$

Equation (8.40) may be written 3 times and solved for h_1, h_2, and h_3 from which b, r_3, and r_6 may be determined. Once again since $\tau = k_5 k_6$, a quadratic equation will result and two values of τ are determined leading to an additional two separate solutions for b, r_3, and r_6.

For three specified positions of a coupler point with three corresponding positions of an input crank, four solutions will result and the designer must select the best of the four or repeat the solution with different values of α, β, and/or θ_{61}.

8.9 COUPLER CURVES WITH GROUND LOOP EQUATIONS AND LEAST SQUARES

The least-squares technique combined with ground loop equations make synthesis of mechanisms to generate specific coupler curves possible with great accuracy. In application of the least-squares technique, Eqs. (8.37) through (8.40) are used to define errors. The error equation is differentiated and set equal to zero to minimize the errors and also to provide for data input. Thus the modified error equations are defined as

$$M_1 = \sum \{e_1[2d_j \cos(\delta_j - \alpha)] + e_2[2d_j \cos(\theta_{1j} - \delta_j)] + e_3 - d_j^2\}^2$$

$$M_2 = \sum \{f_1[2d_j \cos(\delta_j - \alpha)] + f_2[2d_j \cos(\theta_{1j} - \delta_j)] + f_3 - 2\cos(\theta_{1j} - \alpha)\}^2$$

$$M_3 = \sum \{g_5[2d_j \cos(\delta_j - \beta)] + g_6[2d_j \cos(\theta_{6j} - \delta_j)] + g_7 - d_j^2\}^2$$

$$M_4 = \sum \{h_5[2d_j \cos(\delta_j - \beta)] + h_6[2d_j \cos(\theta_{6j} - \delta_j)] + h_7 - 2\cos(\theta_{6j} - \beta)\}^2$$

From $dM_1/de_1 = 0$

$$e_1 \sum 4d_j^2 \cos^2(\delta_j - \alpha) + e_2 \sum 4d_j^2 \cos(\theta_{1j} - \delta_j)\cos(\delta_j - \alpha)$$
$$+ e_3 \sum 2d_j \cos(\delta_j - \alpha) = \sum 2d_j^3 \cos(\delta_j - \alpha) \quad (8.41)$$

From $dM_1/de_2 = 0$

$$e_1 \sum 4d_j^2 \cos(\delta_j - \alpha)\cos(\theta_{1j} - \delta_j) + e_2 \sum 4d_j^2 \cos^2(\theta_{1j} - \delta_j)$$
$$+ e_3 \sum 2d_j \cos(\theta_{1j} - \delta_j) = \sum 2d_j^3 \cos(\theta_{1j} - \delta_j) \quad (8.42)$$

From $dM_1/de_3 = 0$

$$e_1 \sum 2d_j \cos(\delta_j - \alpha) + e_2 \sum 2d_j \cos(\theta_{1j} - \delta_j) + e_3 = \sum d_j^2 \quad (8.43)$$

From $dM_2/df_1 = 0$

$$f_1 \sum 4d_j^2 \cos^2(\delta_j - \alpha) + f_2 \sum 4d_j^2 \cos(\theta_{1j} - \delta_j)\cos(\delta_j - \alpha)$$
$$+ f_3 \sum 2d_j \cos(\delta_j - \alpha) = \sum 4d_j \cos(\theta_{1j} - \alpha)\cos(\delta_j - \alpha) \quad (8.44)$$

From $dM_2/df_2 = 0$

$$f_1 \sum 4d_j^2 \cos(\delta_j - \alpha)\cos(\theta_{1j} - \delta_j) + f_2 \sum 4d_j^2 \cos^2(\theta_{1j} - \delta_j)$$
$$+ f_3 \sum 2d_j \cos(\theta_{1j} - \delta_j) = \sum 4d_j \cos(\theta_{1j} - \delta_j)\cos(\theta_{1j} - \alpha) \quad (8.45)$$

From $dM_2/df_3 = 0$

$$f_1 \sum 2d_j \cos(\delta_j - \alpha) + f_2 \sum 2d_j \cos(\theta_{1j} - \delta_j) + f_3 = \sum 2 \cos(\theta_{1j} - \alpha) \tag{8.46}$$

From $dM_3/dg_5 = 0$

$$g_5 \sum 4d_j^2 \cos^2(\delta_j - \beta) + g_6 \sum 4d_j^2 \cos(\delta_j - \beta) \cos(\theta_{6j} - \delta_j)$$
$$+ g_7 \sum 2d_j \cos(\delta_j - \beta) = \sum 2d_j^3 \cos(\delta_j - \beta) \tag{8.47}$$

From $dM_3/dg_6 = 0$

$$g_5 \sum 4d_j^2 \cos(\delta_j - \beta) \cos(\theta_{6j} - \delta_j) + g_6 \sum d_j^2 \cos^2(\theta_{6j} - \delta_j)$$
$$+ g_7 \sum 2d_j \cos(\theta_{6j} - \delta_j) = \sum 2d_j^3 \cos(\theta_{6j} - \delta_j) \tag{8.48}$$

From $dM_3/dg_7 = 0$

$$g_5 \sum 2d_j \cos(\delta_j - \beta) + g_6 \sum 2d_j \cos(\theta_{6j} - \delta_j) + g_7 = \sum d_j^2 \tag{8.49}$$

From $dM_4/dh_5 = 0$

$$h_5 \sum 4d_j^2 \cos^2(\delta_j - \beta) + h_6 \sum 4d_j^2 \cos(\delta_j - \beta) \cos(\theta_{6j} - \delta_j)$$
$$+ h_7 \sum 2d_j \cos(\delta_j - \beta) = \sum 4d_j \cos(\delta_j - \beta) \cos(\theta_{6j} - \beta) \tag{8.50}$$

From $dM_4/dh_6 = 0$

$$h_5 \sum 4d_j^2 \cos(\delta_j - \beta) \cos(\theta_{6j} - \delta_j) + h_6 \sum 4d_j^2 \cos^2(\theta_{6j} - \delta_j)$$
$$+ h_7 \sum 2d_j \cos(\theta_{6j} - \delta_j) = \sum 4d_j \cos(\theta_{6j} - \delta_j) \cos(\theta_{6j} - \beta) \tag{8.51}$$

From $dM_4/dh_7 = 0$

$$h_5 \sum 2d_j \cos(\delta_j - \beta) + h_6 \sum 2d_j \cos(\theta_{6j} - \delta_j) + h_7 = \sum 2 \cos(\theta_{6j} - \beta) \tag{8.52}$$

Example 8.6. A mechanism is to be designed to create the coupler curve given by d_j and δ_j and at the same time have specified crank angles θ_{1j} in accordance with the following table:

Position	d_j	δ_j	θ_{1j}
1	4.20	38.0	0
2	5.05	40.5	30
3	5.13	47.0	60
4	4.81	56.0	90
5	4.20	66.0	120
6	3.45	75.0	150
7	2.60	83.0	180
8	1.78	86.0	210
9	1.11	77.0	240
10	1.00	47.0	270
11	1.56	32.0	300
12	2.71	36.5	330

Solution

With the selection of $\alpha = \beta = 0$ and $\theta_{61} = 190°$ (which is only a convenient selection), the summations required are formed and Eqs. (8.41) through (8.52) are solved along with Eq. (8.29) resulting in the following four solutions:

	1	2	3	4
a	2.1379	2.1379	5.1617	5.1617
b	1.4293	−3.6075	1.4293	−3.6075
r_1	2.1569	2.1569	−6.1128	−6.1128
r_5	2.4050	2.4050	8.0474	8.0474
r_3	4.3486	1.5897	4.3486	1.5897
r_6	0.8753	−2.1520	0.8753	−2.1520

Solution 1 is shown in Fig. 8.11.

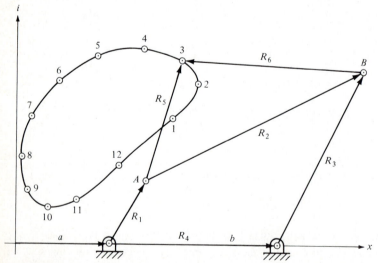

FIGURE 8.11
Solution number 1 for Example 8.6.

8.10 NONLINEAR PROBLEMS

For synthesis of mechanisms with less than four specified positions, the equations are linear. With more than four specified positions or for mechanisms with more than four bars, the equations become nonlinear. The Newton-Rapson method is popular for solution of nonlinear equations. The method uses an iterative procedure to improve an estimate of the solution until the error between the improved estimate and the true value reaches a specified magnitude.

In Fig. 8.12, $f(x)$ is plotted as a function of x for which $f(x) = 0$ is desired. The root is estimated as x_n. A tangent to the curve at x_n will intersect the x axis at x_{n+1}, which is probably a better estimate of the root. The slope of the tangent line is

$$f'(x_n) = \frac{f(x_n)}{x_n - x_{n+1}}$$

then

$$x_{n+1} = x_n - \frac{f(x_n)}{f'(x_n)}$$

Values of the function at the estimated value of x_n, the slope of the tangent line, and the tangent intercept are computed. The value of x_{n+1} is compared with the estimated value of x_n. If the comparison is not within acceptable limits, the computed value of x_{n+1} is used as an estimate and the procedure repeated until the error is within acceptable limits. Care must be taken in making the initial estimate. If x_a is selected as the initial estimate (Fig. 8.12), the calculations may become quite long.

Example 8.7. Freudenstein's equation for a four-bar mechanism is

$$R_1 + R_2 \cos \theta_3 - R_3 \cos \theta_1 = \cos (\theta_1 - \theta_3)$$

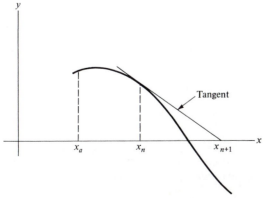

FIGURE 8.12
Newton-Raphson method.

where $R_1 = \dfrac{r_4^2 + r_3^2 + r_1^2 - r_2^2}{2r_1 r_3}$

$R_2 = \dfrac{r_4}{r_1}$

$R_3 = \dfrac{r_4}{r_3}$

With R_1, R_2, and R_3 known, Freudenstein's equation becomes non-linear and it is not possible to solve the equation directly for θ_3 in terms of θ_1. The Newton-Raphson method may be used by assuming a value of θ_3 for a given value of θ_1 and iteration until an acceptable error is reached. To simplify the procedure, the computed value of θ_3 found for the initial value of θ_1 is used as an estimate of θ_3 for the second value of θ_1. The procedure is repeated until all values of θ_1 are consumed. The program NONLIN (App. A) may be used for the problem solution. A rough sketch of the mechanism will provide the initial estimate of θ_3. If the initial estimate of θ_3 is far from the correct value or if the acceptable error is very small, the program may have a very long run time.

PROBLEMS

8.1. Using Freudenstein's equation, synthesize a four-bar mechanism to coordinate crank and rocker displacements as follows:

Position	Crank angle	Rocker angle
1	30°	45°
2	45°	60°
3	60°	90°

8.2. Using Freudenstein's equation, design a function generator for the function $y = e^x$ with $0 \le x \le 1.0$ and precision at $x = 0.067, 0.500, 0.933$.

8.3. Using Freudenstein's equation and the least-squares technique, mechanize the function $y = \sin^2 x$ in accordance with the following table:

Position	$x°$
1	0
2	10
3	20
4	30
5	40
6	50
7	60

8.4. Using the method of components, design a four-bar mechanism to meet the following specifications:

$$r_4 = 10 \qquad r_3 = 8 \text{ at } 60° \qquad \omega_1 = 7 \qquad \omega_3 = 5 \qquad \alpha_1 = \alpha_3 = 0$$

8.5. Using loop equations, design a four-bar mechanism to coordinate the following displacements:

$$\theta_{11} = 20 \qquad \theta_{31} = 30$$
$$\theta_{12} = 40 \qquad \theta_{32} = 60$$
$$\theta_{13} = 60 \qquad \theta_{33} = 90$$

8.6. Design an automated assembly line mechanism which will operate in the horizontal plane and will allow an object at point C_1 to be picked up and moved 2.6 units at an angle of 120° with the horizontal while the object is rotated through 15° to point C_2. Then the object is to be moved to point C_3 located 3.2 units from C_2 at an angle of 140° with the horizontal and the object rotated through 25° from its original orientation. Input crank rotation is to be 20° from position 1 to position 2 and 60° from position 1 to position 3. Output rocker rotation is arbitrary.

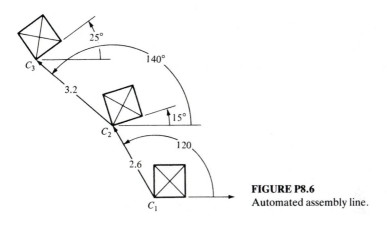

FIGURE P8.6
Automated assembly line.

8.7. Using the ground loop equations, complete design of the four-bar mechanism by finding link lengths r_1, r_3, r_5, and r_6 and vector lengths **a** and **b**. The mechanism is intended to coordinate crank input angles with location of coupler point C.

Position	Input crank position	Point C location
1	60°	$4.90e^{i62}$
2	90°	$4.88e^{i72}$
3	120°	$4.35e^{i82}$

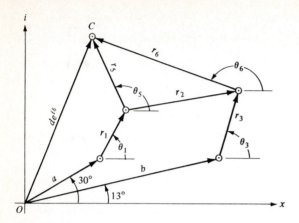

FIGURE P8.7
Synthesis using ground loop equations.

8.8. Using the ground loop equations and least-squares technique, design a four-bar mechanism to generate a coupler curve described by the following table. Use $\alpha = 30°$ and $\beta = 15°$.

Position	d	δ	θ
1	1.84	63.0	0
2	3.31	66.0	30
3	4.00	67.5	60
4	4.15	72.0	90
5	3.87	79.0	120
6	3.40	86.5	150
7	2.80	93.5	180
8	2.10	100.0	210
9	1.80	104.0	240
10	0.79	100.0	270
11	0.51	63.0	300
12	0.85	43.0	330

CHAPTER
9

CAMS

9.1 INTRODUCTION

Programming uniform or intermittent oscillatory motion of a machine component may be accomplished with a cam system. Perhaps the most familiar cam system is the automotive cam whose purpose is to open or close intake and exhaust valves as required by the thermodynamic cycle. The early linotype machine contained many cam systems to move type, cast lead, and even sort the type molds alphabetically. Many modern-day automatic machinery depends upon cams to provide proper timing of the machine components.

A cam profile is the outline of a cam projected onto a plane perpendicular to the cam axis of rotation. This term is misleading when applied to the conical, spherical, cylindrical, or three-dimensional cam systems. The cam profile is that shape of the cam which causes the follower to move in a prescribed manner. Figure 9.1 shows some typical cam profiles.

A cam follower is that link of the cam system which receives the motion prescribed by the cam profile. The cam follower may move in a straight line or swing in an arc. Contact between the cam follower and cam profile may be sliding contact or rolling contact (Fig. 9.2). The axis of motion of the cam follower may or may not pass through the center of rotation of the cam. In many instances the cam follower is held in contact with the cam profile by a spring force. A positive action cam, such as the cylindrical cam of Fig. 9.1, does not need any external force to maintain contact between the follower and the profile.

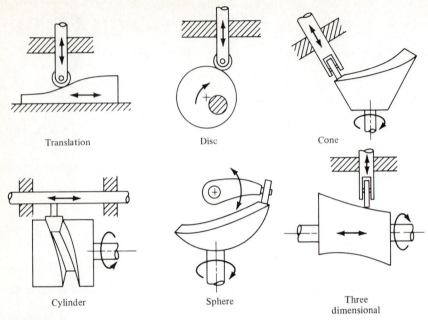

Translation

Disc

Cone

Cylinder

Sphere

Three
dimensional

FIGURE 9.1
Cam types.

The *rise* of a cam is defined as that section of the cam profile, expressed in degrees of cam rotation, during which time the follower moves away from the center of rotation of the cam.

The *fall* of a cam is defined as that section of the cam profile, expressed in degrees of cam rotation, during which time the follower moves toward the center of rotation of the cam.

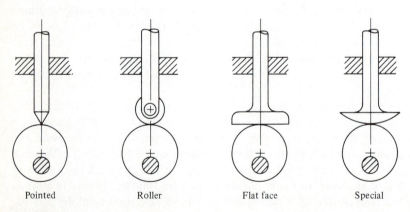

Pointed

Roller

Flat face

Special

FIGURE 9.2
Follower types.

The *dwell* of a cam is defined as that section of the cam profile, expressed in degrees of cam rotation, during which time the follower does not move.

The pressure angle, much like the transmission angle of a mechanism, is a measure of the ease of operation of the cam system. The pressure angle is the angle between the line of action of the cam follower and a normal to the cam profile at the point of contact of the cam follower and the cam profile. A small pressure angle results in very small forces on the follower bearing supports. A large pressure angle will result in very large follower bearing forces and may even result in stall of the machine or breaking of parts.

9.2 CAM DISPLACEMENT DIAGRAMS

The translation cam (Fig. 9.1) forms the basis of all cam systems regardless of the type or style of system. The translation cam consists of a wedge (the cam profile) which is forced to move perpendicular to the motion of the cam follower resulting in displacement of the follower. The translation cam is the same as a displacement diagram for the system. In the case of the disk cam, the displacement diagram which corresponds to the translation cam, is a development of the disk cam profile.

Figure 9.3 shows a disk cam and its related displacement diagram. In the displacement diagram, the vertical axis represents radial motion of

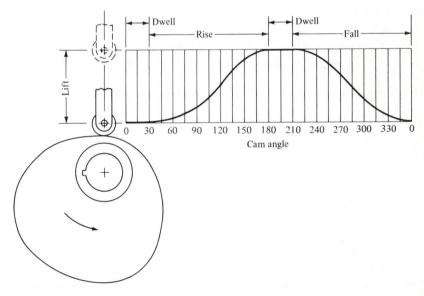

FIGURE 9.3
Cam and related displacement diagram.

the cam follower. The horizontal axis is the developed circumference of the disk cam at its minimum diameter. The horizontal axis of the displacement diagram may be divided into 360 or more segments representing the angular displacement of the cam. Since most cam shafts rotate at constant angular velocity, the horizontal axis of the displacement diagram could also represent time. Note that the displacement diagram is associated with motion of the cam follower and is not directly related to the point of contact of the follower and cam profile.

The shape of the displacement diagram dictates the follower motion. Generally, the shape is determined by requirements of the system. However, the designer, has some control over the pressure angle, velocity, and acceleration of the follower. The system dynamics must be considered before a cam displacement diagram can be accepted. A displacement diagram for a large, slow-moving system may not be acceptable for a high-speed system.

The simplest displacement diagram is one which provides constant velocity to the follower as shown in Fig. 9.4a along with the associated velocity and acceleration diagrams. Since the cam follower which is initially at rest must suddenly move at a constant velocity, the acceleration becomes (theoretically) infinite which results in very large acceleration forces and jerky motion (jerk is the third derivative of displacement). At the top of the rise, the acceleration is again infinite. For this reason, with slow-moving systems, the constant-velocity diagram is modified by including a radius at the start and again at the end of the rise and fall of the follower as shown in Fig. 9.4b.

Rather than constant velocity of the follower and its associated high accelerations and jerks, the constant acceleration or parabolic displacement diagram has been used. With this system, the acceleration and deceleration of the follower is held constant. Parabolic displacement provides for the minimum acceleration possible for a given displacement. However, jerk (derivative of acceleration) is high at the start, middle, and end of the rise or fall of the follower. The constant-acceleration displacement diagram and its associated velocity and acceleration diagrams are shown in Fig. 9.5. Construction of a constant-acceleration curve is shown in Fig. 9.10.

The simple harmonic displacement diagram provides for uniformly changing velocity with some jerk of the cam follower at the start and finish of the simple harmonic motion. Those systems using an eccentric on a rotating shaft to provide follower motion are cams with simple harmonic follower motion.

The projection of a point moving along the circumference of a circle onto the circle diameter as the circle radius moves with constant angular velocity is known as *simple harmonic motion*. In Fig. 9.6, the point P moves along the circumference of a circle with radius equal to $L/2$

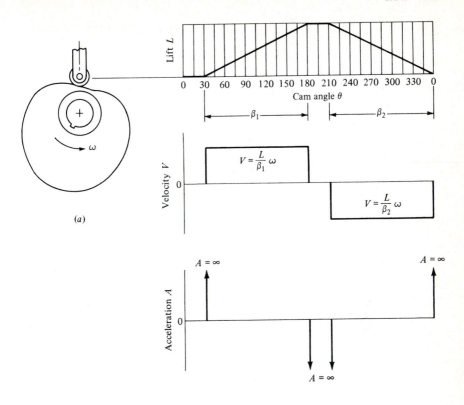

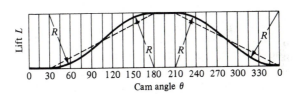

(b) Modified constant-velocity diagram

FIGURE 9.4
Constant follower velocity cam.

(one-half the cam lift). Total angular displacement of the cam is the angle β. Projection of the locus of point P onto corresponding positions of the cam angle results in a simple harmonic displacement diagram. As the generating radius moves through the angle ϕ, the cam rotates through the angle θ and the follower displacement is S. Then

$$S = \frac{L}{2} - \frac{L}{2}\cos\phi = \frac{L}{2(1-\cos\phi)}$$

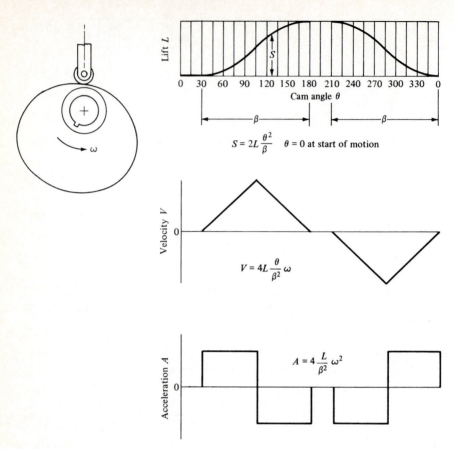

$$S = 2L \frac{\theta^2}{\beta} \qquad \theta = 0 \text{ at start of motion}$$

$$V = 4L \frac{\theta}{\beta^2} \omega$$

$$A = 4 \frac{L}{\beta^2} \omega^2$$

FIGURE 9.5
Constant follower acceleration cam.

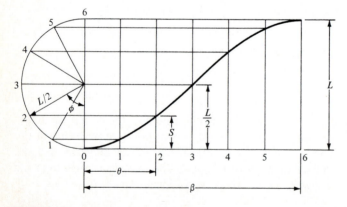

FIGURE 9.6
Simple harmonic curve generation.

With the total cam angle displacement equal to β and the total angle through which the generating radius moves equal to π,

$$\frac{\phi}{\pi} = \frac{\theta}{\beta} \qquad \text{or} \qquad \phi = \frac{\pi\theta}{\beta}$$

then

$$S = \frac{L}{2}\left(1 - \cos\frac{\pi\theta}{\beta}\right)$$

Differentiating the displacement with respect to time and remembering that $d\theta/dt$ is equal to ω gives the velocity as

$$V = \left(\frac{\pi L}{2\beta} \sin\frac{\pi\theta}{\beta}\right)\omega$$

Differentiating the velocity with respect to time gives the acceleration as

$$A = \left(\frac{\pi^2 L}{2\beta^2} \cos\frac{\pi\theta}{\beta}\right)\omega^2$$

A simple harmonic displacement diagram and its associated velocity and acceleration diagrams are shown in Fig. 9.7. Graphical construction technique for simple harmonic motion is shown in Fig. 9.10.

A displacement diagram which provides for jerk-free motion and very good high-speed dynamic characteristics is the cycloidal profile. A *cycloid* is defined as the locus of a point on a circle as the circle rolls along a straight line. Figure 9.8 shows a cycloid generating circle rolling along the vertical axis representing follower displacement. Since the generating circle is to make one complete revolution for the total lift of the cam, its circumference must be equal to the lift L. The radius of the generating circle is then

$$R = \frac{L}{2\pi}$$

As the generating circle rotates through the angle ϕ, the cam rotates through the angle θ. Total cam angle for the complete cycloidal motion is β. Displacement S of the follower will be

$$S = R\phi - R\sin\phi = R(\phi - \sin\phi)$$

Since the generating circle makes one complete revolution for the total rise L,

$$\phi = \frac{2\pi\theta}{\beta}$$

Making substitutions for R and ϕ, the displacement equation becomes

$$S = \frac{L\theta}{S} - \frac{L}{2\pi}\sin\frac{2\pi\theta}{\beta}$$

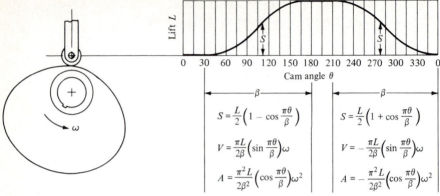

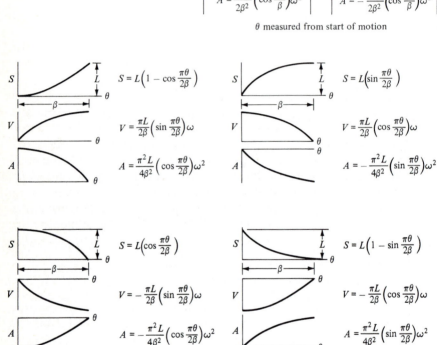

FIGURE 9.7
Simple harmonic follower motion cam. (From M. Kloomok and R. Muffley, "Plate Cam Design," *Prod. Eng.*, February 1955.)

Differentiating the displacement with respect to time gives the velocity and acceleration of the follower as

$$V = \frac{L}{\beta}\left(1 - \cos\frac{2\pi\theta}{\beta}\right)\omega$$

$$A = \frac{2\pi L}{\beta^2}\left(\sin\frac{2\pi\theta}{B}\right)\omega^2$$

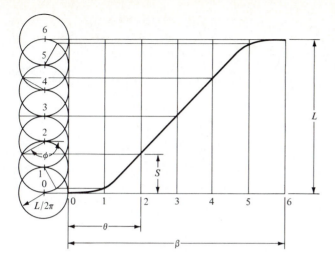

FIGURE 9.8
Generation of a cycloidal curve.

A cycloidal displacement diagram and its associated velocity and acceleration diagrams are shown in Fig. 9.9. Graphical construction of the cycloidal profile is shown in Fig. 9.10.

Special displacement diagrams composed of eighth-order polynomial shapes are used for very high-speed cam rotation.

A displacement diagram may be composed of any of the previous motions or it may be a combination of several of the basic motions.

Example 9.1. It is desired to construct a displacement diagram that will provide simple harmonic motion rise for 90°, constant velocity rise for 65°, simple harmonic motion rise for 45°, 15° of dwell, and cycloidal motion on the return through 145° rotation of the cam shaft. The question to be answered is, What amount of the total 2.5 units lift should be assigned to each of the segments of motion?

Solution
From Fig. 9.7, the velocity at the end of a simple harmonic rise is given by the equation

$$V_1 = \frac{\pi L_1}{2\beta_1} \sin \frac{\pi \theta}{2\beta}$$

At the end of the simple harmonic rise when $\theta = \beta$,

$$V_1 = \frac{\pi L_1}{2\beta_1} \sin \frac{\pi}{2} = \frac{\pi L_1}{2\beta_1}$$

At the start of the constant-velocity motion the follower velocity must be the same as the follower velocity leaving the simple harmonic rise section; otherwise there will be an infinite acceleration and large jerk.

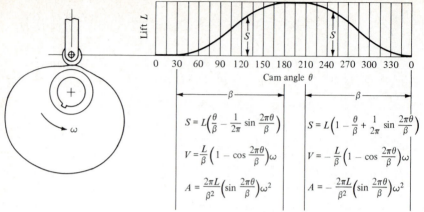

First motion (rise):

$$S = L\left(\frac{\theta}{\beta} - \frac{1}{2\pi} \sin \frac{2\pi\theta}{\beta}\right)$$

$$V = \frac{L}{\beta}\left(1 - \cos \frac{2\pi\theta}{\beta}\right)\omega$$

$$A = \frac{2\pi L}{\beta^2}\left(\sin \frac{2\pi\theta}{\beta}\right)\omega^2$$

Second motion (fall):

$$S = L\left(1 - \frac{\theta}{\beta} + \frac{1}{2\pi} \sin \frac{2\pi\theta}{\beta}\right)$$

$$V = -\frac{L}{\beta}\left(1 - \cos \frac{2\pi\theta}{\beta}\right)\omega$$

$$A = -\frac{2\pi L}{\beta^2}\left(\sin \frac{2\pi\theta}{\beta}\right)\omega^2$$

θ measured from start of motion

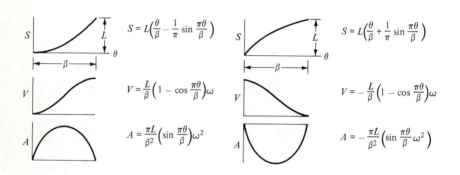

$$S = L\left(\frac{\theta}{\beta} - \frac{1}{\pi} \sin \frac{\pi\theta}{\beta}\right)$$

$$V = \frac{L}{\beta}\left(1 - \cos \frac{\pi\theta}{\beta}\right)\omega$$

$$A = \frac{\pi L}{\beta^2}\left(\sin \frac{\pi\theta}{\beta}\right)\omega^2$$

$$S = L\left(\frac{\theta}{\beta} + \frac{1}{\pi} \sin \frac{\pi\theta}{\beta}\right)$$

$$V = -\frac{L}{\beta}\left(1 - \cos \frac{\pi\theta}{\beta}\right)\omega$$

$$A = -\frac{\pi L}{\beta^2}\left(\sin \frac{\pi\theta}{\beta} \omega^2\right)$$

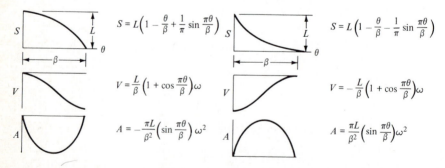

$$S = L\left(1 - \frac{\theta}{\beta} + \frac{1}{\pi} \sin \frac{\pi\theta}{\beta}\right)$$

$$V = \frac{L}{\beta}\left(1 + \cos \frac{\pi\theta}{\beta}\right)\omega$$

$$A = -\frac{\pi L}{\beta^2}\left(\sin \frac{\pi\theta}{\beta}\right)\omega^2$$

$$S = L\left(1 - \frac{\theta}{\beta} - \frac{1}{\pi} \sin \frac{\pi\theta}{\beta}\right)$$

$$V = -\frac{L}{\beta}\left(1 + \cos \frac{\pi\theta}{\beta}\right)\omega$$

$$A = \frac{\pi L}{\beta^2}\left(\sin \frac{\pi\theta}{\beta}\right)\omega^2$$

FIGURE 9.9
Cycloidal follower motion cam. (From M. Kloomok and R. Muffley, "Plate Cam Design," *Prod. Eng.*, February 1955.)

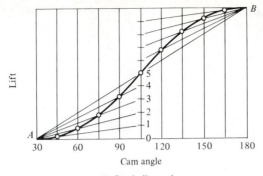

(*a*) Parabolic motion

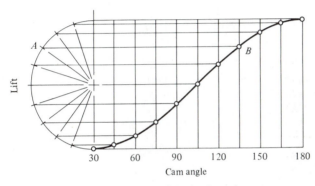

(*b*) Simple harmonic motion

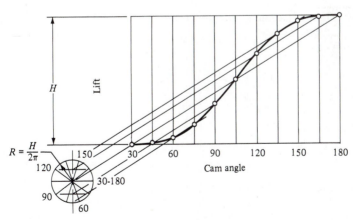

(*c*) Cycloidal motion

FIGURE 9.10
Graphical displacement diagram construction.

Therefore,

$$V_2 = \frac{L_2}{\beta_2} = \frac{\pi L_1}{2\beta_1}$$

or

$$L_2 = L_1 \frac{\pi}{2} \frac{\beta_2}{\beta_1}$$

The follower velocity leaving the constant-velocity section must be equal to the follower velocity entering the last simple harmonic motion section. From Fig. 9.7,

$$V_3 = \frac{\pi L_3}{2\beta_3} \cos \frac{\pi \theta}{2\beta_3} = \frac{L_2}{\beta_2} = \frac{\pi L_1}{2\beta_1}$$

At the start of the last simple harmonic follower motion section $\theta = 0$. Then

$$V_3 = \frac{\pi L_3}{2\beta_3} \cos 0 = \frac{\pi L_3}{2\beta_3} = \frac{L_2}{\beta_2} = \frac{\pi L_1 \beta_2}{2\beta_1}$$

or

$$L_3 = \frac{\pi L_1}{2\beta_1}$$

Since the total lift of the cam is to be equal to 2.5,

$$L_1 + L_2 + L_3 = 2.5$$

or

$$L_1\left(1 + \frac{\pi}{2} \frac{\beta_2}{\beta_1} + \frac{\pi\beta_2}{2\beta_1}\right) = 2.5$$

$$L_1\left(1 + \frac{\pi}{2} \frac{65}{90} + \frac{\pi}{2} \frac{45}{90}\right) = 2.5$$

Then $L_1 = 0.856$ $L_2 = 0.971$ $L_3 = 0.673$

The completed displacement diagram is shown in Fig. 9.11.

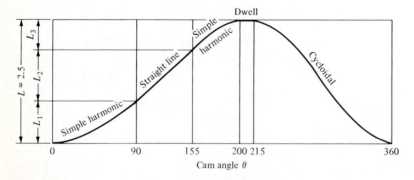

FIGURE 9.11
Cam displacement diagram.

Other cam profile curves may be used to provide the required motion depending upon the motion requirements. An eighth-order polynomial curve used for a displacement diagram will result in peak acceleration and pressure angles intermediate between the cycloidal and the harmonic curves and a nonsymmetrical acceleration curve.

9.3 CAM PROFILE LAYOUT

A cam profile may be developed graphically, analytically, or by computer plot. However the profile is developed, it serves a very useful purpose in determination of space limitations in the machine.

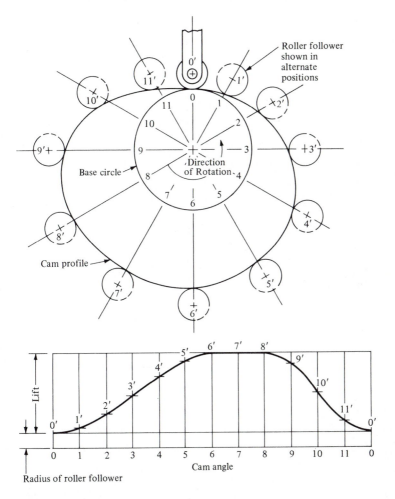

FIGURE 9.12
In-line roller follower cam profile development.

The displacement diagram must be completed and the cam system decided upon before the cam profile can be developed. The lowest position of the cam follower determines the base circle radius. The circumference of the base circle represents the horizontal coordinate of the displacement diagram. In the case of a roller follower, the vertical axis of the displacement diagram represents the position of the center of the roller follower rather than the point of contact between the roller follower and the cam profile.

In Fig. 9.12 an in-line roller follower cam is to be developed. *In-line* simply means that the line of action of the cam follower passes through

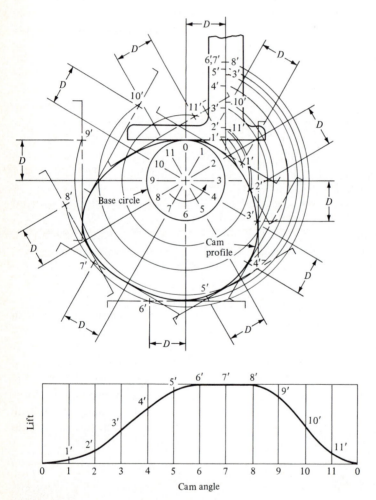

FIGURE 9.13
Offset flat-faced follower cam profile development.

the center of rotation of the cam. An *offset* roller follower system would be one in which the line of action of the follower does not pass through the center of rotation of the cam. The base circle is divided into 12 equal segments for simplicity of illustration. In practice 12 segments is generally not sufficient for industrial accuracy. The base circle segments are numbered in a sense opposite to that of the cam direction of rotation. The displacement diagram is constructed with 12 equal segments representing the 12 segments of the base circle. With the desired displacement diagram constructed, the distances $1 - 1'$, $2 - 2'$, $3 - 3'$, etc., are laid off from the base circle along the corresponding radii. In this manner, the location of the center of the roller follower at each position is determined. Circles equal in diameter to the roller follower are drawn and the cam profile developed as a smooth curve tangent to the roller follower circles. Note that the point of contact between the roller follower and the cam profile is generally not on the radial line of the base circle.

Figure 9.13 shows the graphical layout of an offset, flatfoot, follower cam system. The base circle is divided into 12 equal segments for simplicity of illustration and is numbered in a sense opposite to the sense of rotation of the cam. The displacement diagram is divided into 12 equal segments corresponding to the base circle radial lines. The distances $1 - 1'$, $2 - 2'$, $3 - 3'$, etc., are laid off from the zero position of the follower. At the intersection of a circular arc centered at the cam center of rotation through follower displaced position 1′ with a line parallel to radial line 1 but offset a distance D a line is drawn perpendicular to radial line 1. This line represents the face of the flatfoot follower. The procedure is repeated for all 12 positions, and the cam profile consists of a smooth curve tangent to all lines representing the foot of the follower.

9.4 PRESSURE ANGLE

The pressure angle of a cam system is the angle between the line of action of the follower and a normal to the cam surface at the point of contact. A large pressure angle will result in very large bearing forces on the follower system. A small pressure angle will result in a larger cam system. In general the pressure angle should not exceed 30 or 35° for most systems.

The pressure angle on the rise section of the cam profile is of most importance. The pressure angle exists on the fall section of the cam profile but is of little significance as long as the follower can stay in contact with the profile. The pressure angle for systems with flat-faced followers is of very little importance.

The pressure angle can be made smaller by

1. Increasing the base circle diameter which results in a larger cam system.

2. Increasing the angular displacement devoted to rise of the follower. This is sometimes not possible because of the demands of the design.

3. Decreasing the follower offset. The follower may be offset to either side of the cam centerline but is best offset to the side with oncoming motion in order to reduce the pressure angle on the rise motion.

In the offset roller follower system shown in Fig. 9.14, the pressure angle is α. Rotation of the cam is clockwise. The line of action of the follower is offset from the center of rotation of the cam by the distance e. The rise of the follower is to take place through the angular displacement β. The cam surface is shown as well as the follower pitch surface which is offset from the cam surface a distance equal to the radius of the follower roller, r_r. The vertical distance from the cam center of rotation to the center of the roller follower is equal to the pitch base circle radius) plus the prescribed follower displacement. The distance from the center of rotation of the cam and the center of the roller follower is given by R.

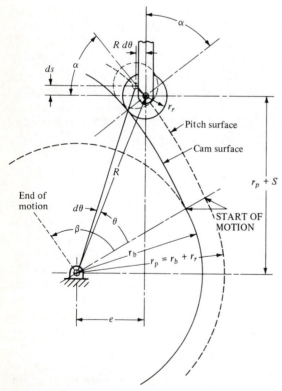

FIGURE 9.14
Cam follower pressure angle.

$$R = \sqrt{e^2 + (r_p + S)^2} \qquad (9.1)$$

With a small angular displacement of the cam $(d\theta)$, the follower will rise a distance ds. With very little error, the corresponding horizontal displacement of the cam follower center will be $R\, d\theta$. Then the pressure angle $\alpha = \arctan{(ds/R\, d\theta)}$.

Example 9.2. A cam system requires the offset follower to rise a distance of 30 units with simple harmonic motion while the cam rotates through 90°. Using a base circle radius of 15 and a roller follower radius of 5 units with an offset of 10 units, determine the maximum pressure angle and its location.

Solution
From Fig. 9.7

$$S = \frac{L}{2}\left(1 - \cos\frac{\pi\theta}{\beta}\right)$$

$$\frac{ds}{d\theta} = \frac{\pi L}{2\beta} \sin\frac{\pi\theta}{\beta}$$

The solution may be accomplished by constructing a table of values and solving for the pressure angle at various cam angles. It is also possible to solve for the pressure angle using the program CAMPA (CAM Pressure Angle, App. A). Output from the program is as follows:

SIMPLE HARMONIC FOLLOWER MOTION
CAM ANGLE FOR TOTAL MOTION, DEGREES   90
FOLLOWER LIFT   30
PITCH BASE CIRCLE RADIUS   20
FOLLOWER OFFSET   10

CAM ANGLE	PRESSURE ANGLE
0	0.00
10	2.59
20	4.41
30	5.18
40	5.08
50	4.43
60	3.48
70	2.37
80	1.20
90	0.00

9.5   ROLLER FOLLOWER DIAMETER

Diameter of the roller follower has a pronounced influence on the cam profile. If the roller is excessively large, the profile may become pointed or even worse may contain loops, which is very unsatisfactory. In the limit the radius of the roller follower must be equal to or preferably smaller than the radius of curvature of the cam profile.

In Fig. 9.15, the radius of the roller follower is r_r and the radius of curvature of the cam profile is ρ. From the differential calculus, the radius of curvature of a curve is given by

$$\rho = \pm \frac{[R^2 + (dR/d\theta)^2]^{3/2}}{R^2 + 2(dR/d\theta)^2 - r(d^2R/d\theta^2)}$$

where if the equation is negative, the curvature of surface is concave, and if the equation is positive, the curvature of surface is convex.

$$R = r_p + S = r_p + f(\theta) \qquad \text{(Fig. 9.15)}$$

$$\frac{dR}{d\theta} = \frac{dS}{d\theta} = \frac{dS}{dt}\frac{dt}{d\theta} = \frac{dS/dt}{d\theta/dt} = \frac{dS/dt}{\omega}$$

$$\frac{d^2R}{d\theta^2} = \frac{(d\theta/dt)(d^2S/d\theta^2) - (dS/dt)(d^2\theta/dt^2)}{d\theta/dt}$$

Since ω is a constant, $\alpha = 0$ and $d^2R/d\theta^2 = (dS/dt)/\omega^2$.

At $\theta = 0$, $S = 0$, and $dS/d\theta = 0$ the radius of curvature of the cam profile is a minimum. With these substitutions:

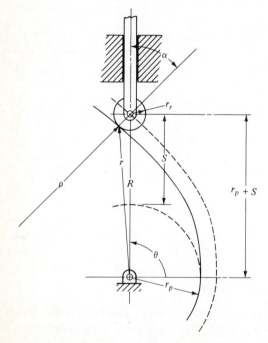

FIGURE 9.15
Radius of roller follower.

$$\rho_{\min} = \pm \frac{[(r_p + S)^2 + (dS/dt)^2/\omega^2]^{3/2}}{(r_p + S)^2 + 2(dS/dt)^2/\omega^2 - (r_p + S)(dS/dt)/\omega^2} \quad (9.2)$$

In the case of a flat-faced follower, the minimum radius of curvature of the cam profile should not become zero and is given by

$$\rho_{\min} = (r_p + S) + \frac{d^2S/dt^2}{\omega^2} > 0 \quad (9.3)$$

9.6   CAM PROFILE PRODUCTION

For cams of very low precision, manufacturing procedure consists of careful layout of the profile on a plate and sawing followed by filing to the finished profile.

Cams requiring greater precision are manufactured using numerically controlled cutters or grinders. In such cases it is necessary to know the coordinates of points on the profile. With reference to Fig. 9.14, the cam pressure angle α along with the base circle radius r_b, the roller follower radius r_r, and the displacement S as a function of cam angle θ are all known. Then the coordinates of any point on the cam profile may be found from

$$r_x = R \cos \theta - r_r \cos \alpha$$
$$\quad (9.4)$$
$$r_y = R \sin \theta - r_r \sin \alpha$$

where $R = \sqrt{c^2 + (r_p + S)^2}$ and $S = f(\theta)$. Angle θ is measured from the start of the rise of the follower.

These equations may be used to generate the cam profile. If the cam is to be cut with an end mill on a numerical control (NC) machine, the path of the center of the end mill is given by

$$R_x = R \cos \theta$$
$$R_y = R \sin \theta$$

where $R = \sqrt{c^2 + (r_p + S)^2}$
$\quad r_r =$ radius of end mill cutter
$\quad S = f(\theta)$

Angle θ is measured from the start of the rise of the follower.

PROBLEMS

Problems 9.1, 9.2, and 9.3 are to be solved using the following displacement requirements of lift versus cam angle.

Cam angle	Lift	Cam angle	Lift	Cam angle	Lift	Cam angle	Lift
0	0.00	100	1.25	190	2.00	280	0.95
10	0.01	110	1.42	200	2.00	290	0.70
20	0.04	120	1.59	210	2.00	300	0.41
30	0.10	130	1.70	220	2.00	310	0.22
40	0.18	140	1.80	230	1.96	320	0.10
50	0.29	150	1.90	240	1.90	330	0.05
60	0.42	160	1.97	250	1.73	340	0.00
70	0.61	170	1.99	260	1.50	350	0.00
80	0.82	180	2.00	270	1.21	360	0.00
90	1.00						

9.1. Lay out a cam profile for a flat-faced in-line follower and determine graphically the minimum length of the follower face. Show the direction of cam rotation. Base circle diameter is 3.0.

9.2. Lay out a cam profile for an in-line roller follower and determine graphically the maximum pressure angle. Roller follower diameter is 0.50. Base circle diameter is 3.00. Show direction of rotation of the cam.

9.3. Lay out a cam profile for an offset roller follower and determine graphically the maximum pressure angle. Roller follower diameter is 0.50. Base circle diameter is 3.00. Follower centerline is offset 0.75 to the right and cam rotation is counterclockwise.

9.4. Lay out a disk cam to give the swing angles of the follower in accordance with the following:

Cam angle	0	30	60	90 to 180	210	240	270 to 360
Swing angle	0	5	10	15 15	10	5	0 0

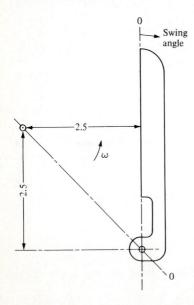

FIGURE P9.4
Rotating follower cam profile development.

9.5. A cam displacement diagram is required to provide a total lift of 1.78 during 85° rotation of the cam. Motion is to be cycloidal for 20°, constant velocity for 40°, and simple harmonic for 25°. Determine the amount of lift devoted to each segment of the rise curve.

9.6. Lay out a cam displacement diagram for a disk cam that will give 0.85 rise with simple harmonic motion for 85° cam rotation, dwell for 15° cam rotation, 1.15 rise with cycloidal motion for 75° cam rotation, dwell for 15° cam rotation, and return with simple harmonic motion.

9.7. A cam system with an in-line roller follower provides a rise of 45 mm with cycloidal motion during 60° of cam rotation. The cam base circle radius is 25 mm.

(*a*) Determine the maximum pressure angle and its location.

(*b*) If the base circle radius is changed to 55 mm, how is the pressure angle altered?

(*c*) With a 55-mm base circle and a 25-mm offset, how is the pressure angle altered?

9.8. A disk cam with simple harmonic motion of a flat-faced follower has a total rise of 1.5 with a cam displacement angle of 120°. The cam has a base circle radius of 0.75. Derive an equation and write a basic program to give the cam profile during rise of the follower. The profile should be in polar form with the angle measured from start of follower motion and the distance from center of rotation of the cam.

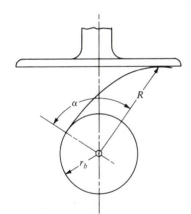

FIGURE P9.8
Harmonic follower motion cam profile development.

CHAPTER

10

SPUR, HELICAL, BEVEL, AND WORM GEARS

10.1 INTRODUCTION

Gears are used for the purpose of transmitting rotary motion between two or more shafts. The rotary motion of the shafts may be at the same angular velocity or it may be at increased or decreased angular velocity of the driver.

If two shafts each contain a cylinder and the two cylinders are in contact, rotation of one shaft will cause the other to rotate in the opposite sense (Fig. 10.1). Such an assembly is known as a friction drive, and the angular velocity of the driven shaft is dependent on friction between the two cylinders. Assuming that there is no slipping of one cylinder on the other, the linear velocity of the two cylinder surfaces must be the same. The ratio of angular velocities of the shafts is inversely proportional to the cylinder's radius (diameter). Thus,

$$V_P = R_2 \omega_2 = R_3 \omega_3$$

or

$$\frac{\omega_3}{\omega_2} = \frac{r_2}{r_3} = \frac{D_2}{D_3}$$

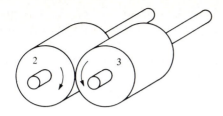

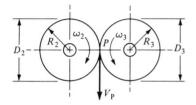

FIGURE 10.1
Friction drive.

In general, discussion of gears uses the gear diameter rather than the gear radius. The diameter is known as the *pitch diameter*.

The cylinders mounted on rotating shafts need not be right circular cylinders (Fig. 10.2). For nonparallel shafts it is possible to use conic sections in contact. Gear teeth may be considered as a means of ensuring

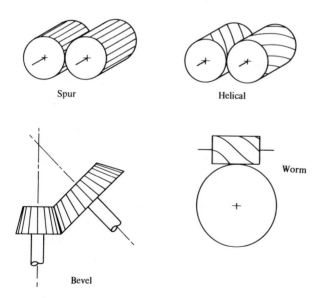

Spur

Helical

Worm

Bevel

FIGURE 10.2
Gear types.

that slipping does not occur. If the gear teeth are cut on the surface of right circular cylinders such that they are parallel to the axis of rotation, the resulting gear is known as a *spur gear*. If the teeth are cut on a right circular cylinder in a helical path, the resulting gear is known as a *helical gear*. If the gear teeth are cut on the surface of a cone, the resulting gear is known as a *bevel gear*. The *spiral bevel* and *hypoid gears* are special cases of the bevel gear. If the gear teeth are cut on the surface of a cylinder in a helical path such that the tooth makes a complete path along the circumference of the cylinder (much like a screw thread), the resulting gear is known as a *worm*.

10.2   GEAR TOOTH PROFILE

Almost any tooth profile will function as a gear profile as long as a proper mating (conjugate) profile is used. However, there is one fundamental principle which must be followed if the angular velocity ratio between shafts is to be constant.

In Fig. 10.3, two profile surfaces are shown in contact with link 2 driving link 3. In all cases, the line of action of the force of one surface which is driving another surface must be perpendicular to the common tangent between the two surfaces at the point of contact. If the angular velocity of the two shafts is to remain constant, the line of action will intersect a line between the centers of the two shafts at the pitch point *P*.

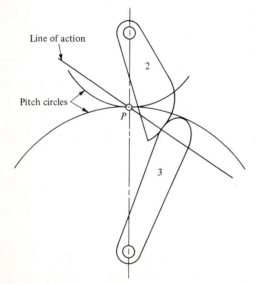

FIGURE 10.3
The pitch point.

The point *P* will be a fixed point on the line of centers if the angular velocity ratio is constant.

Circles drawn through the pitch point are known as the pitch circles of mating gears. Pitch circles represent the cylinder circumferences in friction drives.

Two gear tooth profiles have evolved:

1. *Cycloidal*. The locus of a point on a circle as it rolls on the circumference of another circle. The cycloidal profile is seldom used in machinery today and will not be discussed further.
2. *Involute*. The locus of a point on a taut string as the string is unwrapped from the circumference of a circle. The involute profile is used almost exclusively today.

Figure 10.4 shows a pencil on a string which is wrapped about a base circle. If the string is unwrapped and kept taut, the pencil will describe an involute curve. Two important items are to be noted.

1. The string is always tangent to the base circle.
2. The string is always perpendicular to the involute curve.

If two involute gears were run together as in Fig. 10.5, the force of the driver on the driven gear would be directed along the string which is then known as the line of action. The line of action would always be

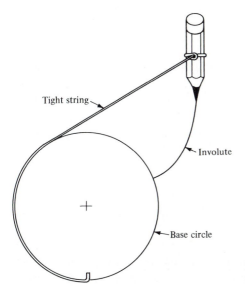

Tight string

Involute

Base circle

FIGURE 10.4
Generation of an involute.

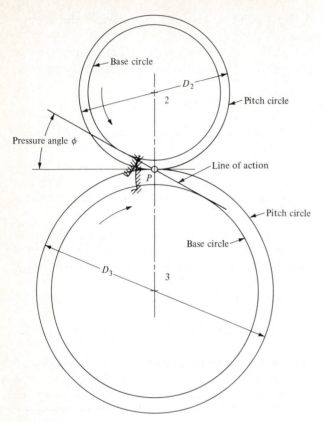

FIGURE 10.5
Line of action.

tangent to the two base circles and would intersect the line of centers at point P, the pitch point. If the center distance of two gears was to be greater than planned for some reason, the line of action would still be tangent to the two base circles. The angle the line of action makes with the common tangent of the two pitch circles would change, but the force transmitted from one gear to the other would be along a line perpendicular to the tooth profile.

The diameter of a gear is the pitch circle diameter or pitch diameter. For a gear alone, it is not possible to measure the pitch diameter. When two gears are placed into contact and the pitch point located, then the pitch diameter of each gear is defined.

The gear tooth extends above and below the pitch circle (Fig. 10.6). The distance from the pitch circle radially outward to the outside diameter of the gear blank is known as the *addendum*, and the outside diameter of the gear blank is the *addendum circle*. The *dedendum* of a

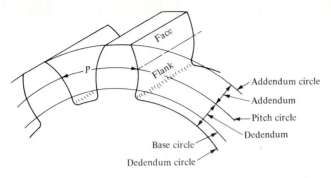

FIGURE 10.6
Spur gear nomenclature.

gear tooth extends radially inward from the pitch circle to the dedendum circle of the gear blank. The dedendum of a gear tooth is larger than the addendum by an amount equal to the clearance. The base circle lies between the pitch circle and the dedendum circle. The whole depth of a gear tooth is equal to the sum of the addendum and dedendum. The working depth is equal to twice the addendum.

The involute profile extends radially outward from the base circle to the addendum circle. The tooth profile between the base circle and the dedendum circle is a straight radial line modified only by the radius of the fillet at the base of the tooth. Radius of the fillet is governed by the gear manufacturing technique.

A gear tooth face is that surface extending from the pitch circle outward to the addendum circle. Note that in considering the strength of gear teeth, the face is considered as the thickness of the blank from which the gear is generated. Although there are two definitions of gear tooth face, no confusion should result if the context of the discussion is known. A gear tooth flank is that surface extending from the pitch circle inward to the dedendum circle. Except in relation to wear and undercutting, the flank of a gear tooth is seldom discussed.

The number of teeth on a gear divided by the diameter of the pitch circle in inches is known as the *diametral pitch P* of the gear. Generally the word diametral is dropped and only the word pitch is used. *Circular pitch p* is the distance from a point on one tooth to the corresponding point on the next adjacent tooth measured along the pitch circle. Circular pitch is never abbreviated to pitch. If the distance is measured along the base circle, it is known as the *base circular pitch* (p_b).

In considering a gear with N teeth and D pitch diameter the diametral pitch is $P = N/D$ and the circular pitch is $p = \pi D/N$. Then

$$N = PD \text{ or } N = \frac{\pi D}{p}$$

and
$$PD = \frac{\pi D}{p}$$

or
$$Pp = \pi$$

In the metric system, the diameter of the pitch circle in millimeters divided by the number of teeth on the gear is known as the *gear module* (*m*).

10.3 STANDARD TOOTH PROFILES

The pressure angle of a gear set is the angle between the line of action of the force and the common tangent to the two pitch circles at the pitch point. Commonly used pressure angles are $14\frac{1}{2}°$, 20°, and 25°. The $14\frac{1}{2}°$ pressure angle is a holdover from the time when gears were cast. The $14\frac{1}{2}°$ pressure angle was popular then because $\sin 14\frac{1}{2}°$ is almost $\frac{1}{4}$ in. With standardization of gear tooth cutters the $14\frac{1}{2}°$ pressure angle was dropped, and it is used today only to replace gears in older machinery. Today the 25° pressure angle is recommended for coarser pitch gears (diametral pitch less than 20). The American Society of Automotive Engineers recommends a 20° pressure angle for general use and 22.5 or 25° pressure angle for aerospace applications.

The commonly used pressure angles have evolved over many years, and a large inventory of gear tooth cutters and generators is available for those pressure angles. Although use of other pressure angles is possible, it is to be discouraged since manufacturing will be costly and replacement of worn gears may not be possible in all locations

Standard dimensions of gear teeth are established by the American Gear Manufacturers Association (AGMA) and, for metric gears, the International Standard ISO 701, the British standard metric, and the German standard metric.

Standard spur gear symbols as identified by the AGMA and ISO 701 are shown in Table 10.1.

The AGMA standard tooth proportions for involute spur gears are given in Table 10.2.

The diametral pitch for commercially available gear cutters with either $14\frac{1}{2}$ or 20° pressure angle is as follows:

From $\frac{1}{2}$ to 4 in increments of $\frac{1}{2}$
4 to 10 in increments of 1
10 to 32 in increments of 2
120 to 200 in increments of 2
Also 32, 36, 40, 42, 48, 50, 64, 72, 80, 96, 120

TABLE 10.1
Standard spur gear symbols

	AGMA	ISO 701
Addendum	a	h_a
Backlach	B	j_t
Base pitch	p_b	p_b
Base radius	R_b	r_b
Center distance	C	a
Circular pitch	p	p
Clearance	c	c
Contact ratio	m_p	ϵ_a
Dedendum	b	h_f
Diametral pitch	P	
Face width	F	b
Length of action	Z	g_a
Module	. . .	m
Number of teeth	N	z
Outside diameter	D_o	d_a
Outside radius	R_o	r_a
Pitch diameter	D	d'
Pitch radius	R	r
Pressure angle	ϕ	α
Tooth thickness	t	s
Whole depth	h_t	h
Working depth	h_k	

TABLE 10.2
AGMA involute spur gear tooth proportions

		Full-depth gears		Stub gears
		Coarse pitch	Fine pitch	
Pressure angle, °	$14\frac{1}{2}$	20 or 25	20	20
Addendum	$1.000/P$	$1.000/P$	$1.000P$	$0.800/P$
Dedendum	$1.157/P$	$1.250/P$	$1.2/P + 0.002$	$1.000/P$
Clearance	$0.157/P$	$0.250/P$	$0.2/P + 0.002$	$0.200/P$
Fillet radius	$0.209/P$	$0.300/P$	None given	$0.304/P$

The metric standard tooth proportions for involute spur gears is given in Table 10.3.

TABLE 10.3
Metric standard tooth proportions

	British standard	German standard
Pressure angle, °	20	20
Addendum	$1.000m$	$1.000m$
Dedendum	$1.250m$	$1.157m$ or $1.167m$

The preferred module for British standard metric gear cutters with 20° pressure angle is as follows:

From 1 to 3 in increments of 0.25
Also 5, 6, 8, 10, 12, 16, 20, 25, 32, 40, 50

A second-choice module for British standard metric gear cutters provides choice of modules between those listed as preferred.

The modules for German standard metric gear cutters with 20° pressure angle is as follows:

From 0.3 to 1.0 in increments of 0.10
      1.0 to 4.0 in increments of 0.25
      4.0 to 7.0 in increments of 0.50
      7.0 to 18.0 in increments of 1.00
    18.0 to 24.0 in increments of 2.00
    24.0 to 45.0 in increments of 3.00
    45.0 to 75.0 in increments of 5.00

Gears manufactured using standard cutters and in which the tooth thickness at the pitch circle is equal to the tooth space width at the pitch circle are known as *standard gears*. In some instances a standard gear may not perform as well as desired and the designer must resort to nonstandard gears. Nonstandard gears are manufactured with the same standard cutters, but the tooth thickness and tooth space at the pitch circle are not equal.

10.4   INTERFERENCE

Two gear teeth will operate properly with the line of action perpendicular to a common tangent at the point of contact. Since the gear tooth is a straight radial line from the base circle to the dedendum circle, if one gear contacts the other inside the base circle, a condition known as interference exists. *Interference* is then defined as that time of contact in which an involute profile is in contact with a noninvolute profile.

Contact between two gears begins when the addendum of one gear crosses the line of action. Contact ends between two gears when the addendum of the other gear crosses the line of action. The length of the contact line extends along the line of action from one addendum circle to the other addendum circle. In order to avoid interference, the addendum of a gear must cross the line of action between its point of tangency with the mating gear base circle and the pitch point. An addendum circle which crosses the line of action outside its point of tangency with the mating gear base circle results in interference.

In Fig. 10.7, gear 2 is driving gear 3. The maximum addendum circle of gear 3 intersects the line of action at point A. The maximum addendum circle of gear 2 intersects the line of action at point B. The distance AB is the length of the contact line. Then AB is the maximum length of the contact line if interference is to be avoided. Then,

$$PA = R_2 \sin \phi$$

$$PB = R_3 \sin \phi$$

and $AB = PA + PB = C \sin \phi = $ maximum length of contact where C is the center distance and is equal to $R_2 + R_3$.

It is not always possible nor desirable to use the maximum permissible addendum circle. In the case of standard gear tooth profiles, the addendum is something less than the maximum possible. In Fig. 10.8, if the addendum is less than the maximum permissible, the length of the contact line Z is CD. Thus,

$$Z = CD = CP + PD$$

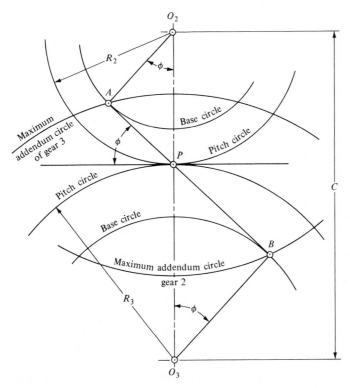

FIGURE 10.7
Maximum length of contact line.

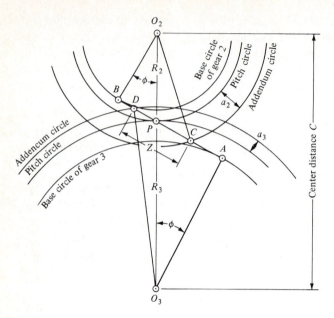

FIGURE 10.8
Length of contact line.

and
$$CP = BC - BP = \sqrt{(O_2C)^2 - (O_2B)^2} - R_2 \sin \phi$$
$$PD = AD - AP = \sqrt{(O_3D)^2 - (O_3A)^2} - R_3 \sin \phi$$

where R_2 = pitch radius of gear 2
  O_2C = addendum radius of gear 2 = $R_2 + a$
  O_2B = base circle radius of gear 2 = $R_2 \cos \phi$
  R_3 = pitch radius of gear 3
  O_3D = addendum radius of gear 3 = $R_3 + a$
  O_3A = base circle radius of gear 3 = $R_3 \cos \phi$
  $C = R_2 + R_3$ = center distance

Then
$$Z = CD = \sqrt{(O_2C)^2 - (O_2B)^2} + \sqrt{(O_3D)^2 - (O_3A)^2} - C \sin \phi$$

10.5   CONTACT RATIO

The contact ratio m is a measure of quality of performance of a gear set. The *contact ratio* is defined as the number of base circular pitches contained in the length of the contact line. For gear teeth with involute profile, the contact line actually represents a development of the circumference of the base circle. Dividing the length of the contact line by the base circular pitch will determine the number of teeth which could be in contact at one time. Therefore,

$$m_p = \frac{\text{length of contact line}}{\text{base circular pitch}}$$

$$\text{Base circular pitch} = \frac{\text{number of teeth on gear}}{\text{circumference of base circle}}$$

The contact ratio should be as high as possible. A high-contact ratio will result in gear teeth which operate smoothly. If the contact ratio is less than 1, the gear set will not function continuously. A contact ratio less than 1.4 is not recommended.

10.6 MINIMUM NUMBER OF TEETH

As the size of a gear becomes smaller, the possibility of interference increases. In order to determine the minimum number of teeth permissible on a gear so that no interference will occur consider the gear in Fig. 10.9 which is shown in contact with a rack. A *rack* is a gear with infinite radius of the pitch circle.

Addendum height of the rack is given as k/P in which k is a factor defining full depth or stub teeth and P is the diametral pitch of the gear. Then,

$$AB = \frac{k}{P} = PA \sin \phi = R \sin \phi \sin \phi = R \sin^2 \phi$$

With the definition of diametral pitch $R = N/2P$

$$\frac{k}{P} = \frac{N}{2P} \sin^2 \phi$$

or

$$N = \frac{2k}{\sin^2 \phi}$$

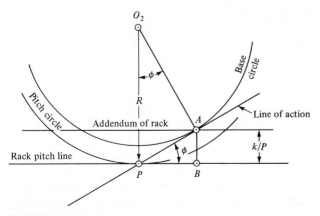

FIGURE 10.9
Rack and gear in contact.

TABLE 10.4
Minimum number of teeth to run with a rack

Pressure angle, °	$14\frac{1}{2}$	20	20	25	25
Tooth depth	Full	Full	Stub	Full	Stub
Minimum number of teeth	32	18	14	12	9

$$\text{where } k = \begin{cases} 1.0 & \text{for full-depth gear teeth} \\ 0.8 & \text{for stub gear teeth} \end{cases}$$

If a gear will run with a rack without interference, it will also run with a gear of less radius than a rack without interference. Table 10.4 gives the minimum number of teeth allowed on a gear if it is to run without interference with a rack.

10.7 BACKLASH

Backlash is the amount the tooth space is larger than the tooth width. Backlash is necessary to ensure that the gear teeth mesh without binding. If the tooth space and the tooth width were designed to be the same amount and the center distance was a little less than intended, the gear teeth would bind.

A standard tooth thickness at the pitch circle is equal to one-half the circular pitch. Thus the thickness at one radius is known. In Fig. 10.10, the tooth thickness at radius R is equal to t and one-half the tooth thickness is equal to $t/2$. It is desired to calculate the tooth thickness t_A at radius R_A. Since the tooth profile is involute, the distance AB must be

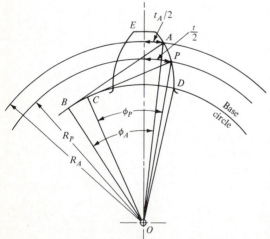

FIGURE 10.10
Involute tooth thickness.

equal to the arc distance DB. Then,

$$\text{Angle } DOB = \frac{\text{arc } DB}{OD} = \frac{AB}{OB} = \tan \phi_A$$

$$\text{Angle } DOC = \frac{\text{arc } DC}{OD} = \frac{PC}{OC} = \tan \phi_P$$

and $\text{Angle } DOA = \text{angle } DOB - \phi_A = \tan \phi_a - \phi_A = \text{inv } \phi_A$

$\text{Angle } DOP = \text{angle } DOC - \phi_P = \tan \phi_P - \phi_P = \text{inv } \phi_P$

Now

$$\text{Angle } DOE = \text{angle } DOA + \frac{\frac{1}{2}t_A}{R_A} = \text{inv } \phi_A + \frac{\frac{1}{2}t_A}{R_A}$$

$$\text{Angle } DOE = \text{angle } DOP + \frac{\frac{1}{2}t}{R} = \text{inv } \phi_P + \frac{\frac{1}{2}t}{R}$$

Equating the two expressions for angle DOE provides

$$\text{inv } \phi_A + \frac{\frac{1}{2}t_A}{R_A} = \text{inv } \phi_P + \frac{\frac{1}{2}t}{R}$$

or $$t_A = 2R_A\left(\text{inv } \phi_P - \text{inv } \phi_A + \frac{\frac{1}{2}t}{R}\right) \qquad (10.1)$$

where t_A = thickness of gear tooth at radius R_A

ϕ_P = pressure angle at pitch circle radius R

ϕ_A = pressure angle at radius R_A

t = one-half circular pitch of gear

In Fig. 10.11, a line from any point on the involute profile and perpendicular to the profile will be tangent to the base circle. The base circle radius is given by

$$R_b = R \cos \phi = R_A \cos \phi_A$$

Then

$$\phi_A = \arccos\left(\frac{R}{R_A} \cos \phi\right)$$

Figure 10.12 shows two gears cut with standard teeth and in contact at the pitch point. The line of action makes the angle ϕ (the cutter pressure angle) with the common tangent of the pitch circles at the pitch point. If the center distance is increased a slight amount ΔC, the line of action will remain tangent to the two base circles and the pressure angle will increase a slight amount to the angle ϕ_A. This means that the pitch circles will be increased to operating pitch circles with radii R_2' and R_3'.

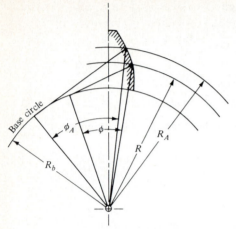

FIGURE 10.11
Operating pressure angle.

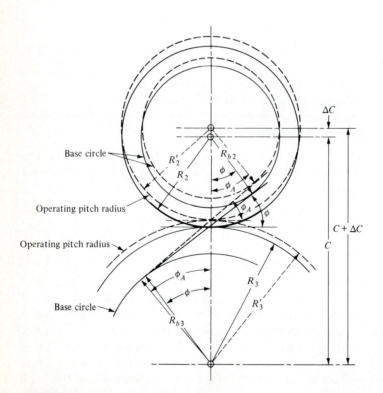

FIGURE 10.12
Extended center distance.

Since the angular velocity ratio is unchanged, the ratio is inversely proportional to the numbers of teeth on the gears or to the operating pitch radii, and the extended center distance is equal to the sum of the operating pitch radii,

$$R_2' = \frac{N_2}{N_2 + N_3}(C + \Delta C)$$

$$R_3' = \frac{N_3}{N_2 + N_3}(C + \Delta C)$$

Using Eq. (10.1) the tooth thicknesses on the operating pitch circles may be calculated as t_2' and t_3'. The backlash shown in Fig. 10.13 is found by adding tooth thicknesses and backlash and equating to the circular pitch on the operating pitch circle.

$$B + t_2' + t_3' = \frac{2\pi R_2'}{N_2} = \frac{2\pi R_3'}{N_3}$$

or

$$B = \frac{2\pi R_2'}{N_2} - t_2' - t_3' \tag{10.2}$$

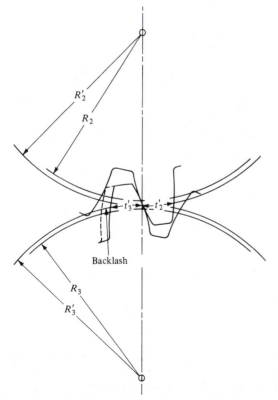

FIGURE 10.13
Backlash.

$$B = \frac{2\pi R_3'}{N_3} - t_2' - t_3'$$

Remembering that, for standard gears,

$$p = \frac{\pi}{P_d} = \frac{2\pi R}{N}$$

or $\qquad p = \pi m/2 \qquad$ for metric gears

$$t_2 = t_3 = \frac{p}{2} = \frac{\pi}{2P_d}$$

or $\qquad t_2 = t_3 = \pi m/2 \qquad$ for metric gears

it is possible to substitute Eq. (10.1) into Eq. (10.2) and to convert the equation to

$$B = (C + \Delta C)(\text{inv } \phi_A - \text{inv } \phi) \qquad (10.3)$$

Recommended values for backlash are listed in the *AGMA Gear Handbook*, Washington, D.C., March 1980.

10.8   INTERNAL GEARS

If the involute profile gear teeth are cut on the inside of a cylinder rather than on the outside surface, an internal gear results. The involute profile will be concave rather than complex. An interesting type of interference is possible with internal gear teeth that is not possible with external teeth. A situation known as *fouling* can occur. Fouling exists when the addendum surface of two mating teeth contact one another. To avoid fouling, the addendum of internal gear teeth face surface is rounded slightly to provide more clearance.

The relationships $P_d = D/N$ or $m = N/D$ hold for internal teeth as well as for external teeth. The line of action is defined in the same manner as for external teeth. Figure 10.14 shows an internal gear mated with a spur gear. Note that when using an internal gear the direction of rotation of the internal gear and the driving gear is the same. With external gears, the direction of rotation of the gears is opposite. Thus when the driver of an external gear is rotating clockwise, the driven external gear will rotate counterclockwise while a driven internal gear will rotate clockwise.

10.9   NONSTANDARD GEARS

In some cases it is not possible to obtain an exact angular velocity ratio using standard spur gears, and the designer may resort to nonstandard gears. It is also possible to improve the load-carrying capacity or to

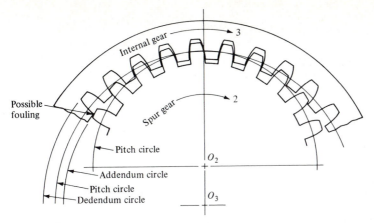

FIGURE 10.14
Internal gear.

reduce the noise generated by using nonstandard gears. The subject of nonstandard gears is rather involved and will not be pursued further other than to mention the various procedures in common use.

One of the major problems of standard gears is undercutting or interference in which the involute profile of one gear runs on the straight radial section of the mating gear. When undercutting will occur, the cutter used in manufacture of the gear will remove a portion of the flank of the gear tooth resulting in a thinner segment at the root of the tooth. At the same time a reduction of the length of contact line and reduction of contact ratio will occur. This problem may be corrected by using an extended center distance technique in which the cutter is withdrawn a slight amount on one or both gears and the center distance is modified slightly, resulting in an increase in the pressure angle. The technique will not work if the angular velocity ratio is 1.0. However, the system does provide for stronger teeth and improved contact ratio.

The long and short addendum technique is one in which the cutter is withdrawn from the pinion an amount equal to that which the cutter is advanced into the gear. This results in gear teeth of proportions different from those given by the standards but will in many cases cure the interference and undercutting problem. In addition the procedure does not alter the pressure angle. The technique works well for 20° pressure angle gears in which the sum of the numbers of teeth is 34 or more.

10.10   HELICAL GEARS

Problems of obtaining specified gear ratios and quieter operation are well handled with helical gears. A helical gear is one in which the gear teeth are not cut parallel to the axis of rotation of the gear but are cut at an

TABLE 10.5
Helical gear nomenclature

	AGMA	ISO 701
Circular pitch in plane of rotation	p	p
Circular pitch in normal plane	p_n	p_n
Helix angle	ψ	β
Pitch in plane of rotation	P	
Pitch in normal plane	P_n	
Module in plane of rotation	. . .	m
Module in normal plane	. . .	m_n
Pressure angle in plane of rotation	ϕ	α
Pressure angle in normal plane	ϕ_n	α_n
Shaft angle	Σ	Σ

angle known as the helix angle. Table 10.5 gives standard nomenclature for helical gears.

Figure 10.15 shows the relationship between the circular pitch p and the normal circular pitch p_n of a helical gear.

$$p_n = p \cos$$

or $$p_n = \pi m \cos \psi \qquad \text{for metric gears} \qquad (10.4)$$

where ψ is the helix angle measured from the axis of rotation.

Since $pP = \pi$ ($p/m = \pi$ for metric gears) and $P = N/D$ ($m = D/N$ for metric gears), substitution in Eq. (10.4) results in

$$D = \frac{N}{P_n \cos \psi} \qquad (10.5)$$

or $$D = \frac{Nm_n}{\cos \psi} \qquad \text{for metric gears} \qquad (10.6)$$

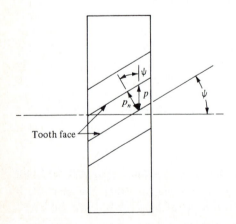

FIGURE 10.15
The helix angle.

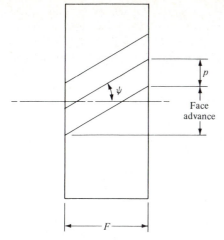

FIGURE 10.16
Face width of a helical gear.

The outside diameter of a helical gear using standard tooth profile may be calculated as follows:

$$D_o = D + 2a$$

where D_o = outside diameter of gear blank
D = pitch diameter = N/P
a = addendum = k/P_n
P_n = pitch normal to the gear tooth = pitch of cutter
P = pitch in plane of rotation = $P_n \cos \psi$

For smooth operation, the face width of a helical gear is made large enough to ensure that the face advance is greater than the circular pitch. The AGMA recommends that a 15 percent margin of safety be provided. Thus the face width F of Fig. 10.16 is given by

$$F > \frac{1.15p}{\tan \psi}$$

10.11 PARALLEL SHAFT HELICAL GEARS

In order for parallel shaft helical gears to mesh and operate properly three conditions must be observed:

1. The gears must have the same helix angle.
2. The gears must be of opposite hand. One gear has a right-hand helix and the other a left-hand helix.
3. The gears must be of the same pitch or module.

Example 10.1. A gear set is to be cut with a 10-pitch 20° pressure angle cutter using stub teeth. The assembly is to provide a 3.6:1 angular velocity ratio and have a 6.875-in center distance. It is necessary to specify the numbers of teeth on each gear and each gear blank outside diameter and face width.

Solution
Since the angular velocity ratio is 3.6:1, one gear will have 3.6 times more teeth than the other.

$$N_3 = 3.6N_2$$

Center distance will be equal to the sum of the gear radii.

$$C = R_2 + R_3 = \frac{N_2 + N_3}{2P} = \frac{N_2 + 3.6N_2}{2P} = \frac{2.3N_2}{P}$$

Since the center distance is to be 6.875,

$$P = \frac{2.3N_2}{6.875} = 0.3346N_2$$

It is now necessary to construct a table and calculate the pitch which will result from a whole number of teeth. Recall that the pitch to be calculated is the pitch in the plane of rotation of the gear. By selecting a pitch of 10 which is the pitch of the cutter, spur gears are being considered.

N_2	N_3	P	Comments
29.89	107.61	10	Unsatisfactory spur gears
28	100.8	. . .	Unsatisfactory N_3
27	97.2	. . .	Unsatisfactory N_3
26	93.6	. . .	Unsatisfactory N_3
25	90.0	8.3638	Satisfactory

The table is stopped when a whole number of teeth is found for each gear. Then

$$\cos \psi = \frac{8.3638}{10} \quad \text{and} \quad \psi = 33.24°$$

Outside diameter of the gear blanks will be

$$D_{o2} = D_2 + 2a = \frac{25}{8.3638} + 2\frac{0.8}{10} = 3.1491$$

$$D_{o3} = D_3 + 2a = \frac{90}{8.3638} + 2\frac{0.8}{10} = 10.9207$$

Minimum face width (thickness of gear blank) will be

$$F = \frac{1.15p}{\tan \psi} = \frac{1/15\pi}{P_n \tan 33.24}$$

$$F = \frac{1.15\pi}{8.3638 \tan 33.24} = 0.6591$$

The recommendation is to use helical gears with 25 teeth and 90 teeth. The helix angle will be 33.24°. Gear 2 will have an outside diameter of 3.1491 in, and gear 3 will have an outside diameter of 10.9207 in. Both gears will have a minimum thickness of 0.6591 in. Since 0.6591 is not a standard material thickness, it is not really necessary to take a thicker standard piece of material and cut it down to size unless weight or space is a problem.

Example 10.2. A gear set is to be cut with a German metric module 2.5 cutter with 20° pressure angle. The gear set is to provide 3.6:1 angular velocity ratio with a 175.0-mm center distance. It is necessary to recommend the numbers of teeth on each gear and dimensions of the gear blanks.

Solution

$$\text{Center distance} = R_2 + R_3 = \tfrac{1}{2}(mN_2 + mN_3)$$

With 3.6:1 angular velocity ratio $N_3 = 3.6N_2$. Then,

$$C = \tfrac{1}{2}(4.6N_2)m = 175$$

or

$$N_2m = 76.0870$$

As in Example 10.1 a table is constructed.

N_2	N_3	m	Comments
30.44	109.6	2.5	Unsatisfactory spur gears
30	108	2.536	Satisfactory

Then

$$\cos \psi = \frac{2.5}{2.536} \quad \text{and} \quad \psi = 9.697°$$

Outside diameters of the gear blanks become

$$D_{o2} = N_2m + 2m_n = (30)(2.536) + 2(2.5) = 81.080 \text{ mm}$$

$$D_{o3} = N_3m + 2m_n = (108)(2.536) + 2(2.5) = 278.888 \text{ mm}$$

Face width (thickness of the gear blank) becomes

$$F = \frac{1.15p}{\tan \psi} = \frac{1.15\pi m}{\tan 9.697} = \frac{(1.15)(\pi)(2.536)}{\tan 9.697} = 78.03 \text{ mm}$$

10.12 NONPARALLEL SHAFT HELICAL GEARS

Helical gears may be used to connect shafts which are not parallel but are inclined to one another by the angle Σ. As a consequence of the geometry of the system, there will be point contact between two mating teeth and a sliding action and rapid wear will result. For this reason, nonparallel shaft helical gears are restricted to use in systems requiring small amounts of power. The involute tooth profile is used in most instances. However, with care in manufacturing, line contact can be obtained but at the sacrifice of the involute profile.

In order for two nonparallel shaft helical gears to function properly they

1. Must have the same normal pitch
2. Need not have the same helix angles
3. Need not have the same pitch in the plane of rotation
4. May be of the same or opposite hand

The angle between the two shafts will be equal to the sum or the difference of the helix angles.

$$\Sigma = \begin{cases} \psi_2 + \psi_3 & \text{for gears of the same hand} \\ \psi_2 - \psi_3 & \text{for gears of opposite hand} \end{cases}$$

Example 10.3. A lightly loaded pair of shafts intersecting at a 55° angle are to be connected by helical gears giving a 2.5:1 velocity ratio. Gear 2 has a normal pitch of 8, a pitch diameter of 6.7803 in the plane of rotation, and a helix angle ψ_2 of 32°. It is desired to know the numbers of teeth on each gear and the diameter and helix angle of the mating gear if the gears are of the same hand.

Solution

$$\Sigma = 55 = \psi_2 + \psi_3$$

$$\psi_3 = 55 - 32 = 23°$$

$$\frac{\omega_2}{\omega_3} = \frac{N_3}{N_2} = \frac{D_3 \cos \psi_3}{D_2 \cos \psi_2}$$

$$D_3 = \frac{D_2 \cos \psi_2}{\cos \psi_3} \frac{N_3}{N_2} = \frac{6.7803 \cos 32}{\cos 23} 2.5 = 15.6165$$

$$N_2 = P_n D_2 \cos \psi_2 = (8)(6.7803) \cos 32 = 46$$

$$N_3 = P_n D_3 \cos \psi_3 = (8)(15.6165) \cos 23 = 115$$

10.13   WORM GEARS

If the gear tooth of a helical gear makes a complete revolution about the pitch surface, it is known as a worm. The mating gear is known as the worm gear or worm wheel. The worm and worm gear system is generally used with nonintersecting shafts oriented at 90°. Although it is possible to use something less than 90°, it is seldom done. Additional nomenclature associated with worm gearing is given in Table 10.6.

TABLE 10.6
Worm gear nomenclature

	AGMA	ISO 701
Axial pitch (worm)	p	p
Circular pitch (gear/wheel)	p_x	p_x
Helix angle	ψ	β
Lead	L	p_z
Lead angle	λ	γ

A worm and worm gear with shafts at right angles must satisfy the following conditions in order to operate properly.

1. The lead angle λ of the worm must equal the helix angle ψ of the worm gear.
2. Axial pitch p of the worm must equal the circular pitch p_x of the worm gear.

The lead of a worm is defined as the axial distance that a point on the tooth will advance in one revolution of the worm. The axial pitch and lead are related as shown in Fig. 10.17.

$$L = p_x N$$

where N is the number of teeth wrapped on the pitch surface.

A circumferential development of one tooth on the worm defines the lead angle and helix angle of a worm as shown in Fig. 10.18.

$$\tan \lambda = \frac{L}{\pi D} = \frac{1}{\tan \psi}$$

The pitch diameter of a worm wheel is given by

$$D = \frac{pN}{\pi}$$

where p is the circular pitch of the worm gear or axial pitch of the worm.

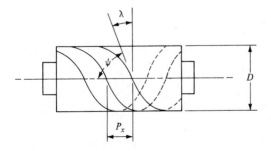

FIGURE 10.17
Worm nomenclature.

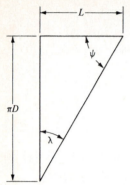

FIGURE 10.18
Worm lead and helix angles.

Example 10.4. A quadruple thread worm is driving a 160-tooth worm wheel with the shafts at 90°. Circular pitch of the worm gear is 1.125 in. Pitch diameter of the worm is 2.75 in. It is desired to know the worm lead angle, helix angle of the worm gear, and the distance between shaft centers. The assembly is shown in Fig. 10.19.

Solution
Lead of the worm is given by

$$L = pN$$

where $p = p_x$ is the circular pitch of the worm.

$$L = (1.125)(4) = 4.5$$

The lead angle is

$$\tan \lambda = \frac{L}{\pi D} = \frac{4.5}{(\pi)(2.75)} = 0.5209$$

$$\lambda = 27.51° = \text{helix angle of the worm gear}$$

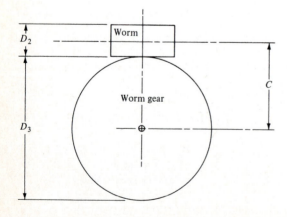

FIGURE 10.19
Worm and gear assembly.

Diameter of the worm gear is

$$D = \frac{pN}{\pi} = \frac{(1.125)(160)}{\pi} = 57.2958 \text{ in}$$

Center distance will be

$$C = \tfrac{1}{2}(D_2 + D_3) = \tfrac{1}{2}(2.75 + 57.2958) = 30.0229 \text{ in}$$

Angular velocity ratio will be

$$\frac{\omega_2}{\omega_3} = \frac{N_3}{N_2} = \frac{160}{4} = 40$$

10.14   BEVEL GEARS

If the gear teeth are cut on the surfaces of cones rather than cylinders, a bevel gear results as shown in Fig. 10.20. Nomenclature associated with bevel gears is given in Table 10.7.

The angular velocity ratio of bevel gears is inversely proportional to the numbers of teeth or to the pitch diameters. Pitch diameter is taken as the base diameter of the pitch cone.

$$\frac{\omega_2}{\omega_3} = \frac{N_3}{N_2} = \frac{D_3}{D_2}$$

With reference to Fig. 10.20, relationships between the cone angles, shaft angle, and angular velocity ratio may be developed.

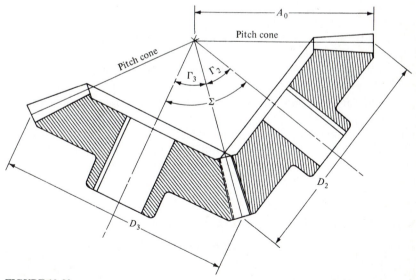

FIGURE 10.20
Typical bevel gears.

TABLE 10.7
Bevel gear nomenclature

	AGMA	ISO 701
Addendum angle	α	θ_a
Cone distance	A_0	R
Dedendum angle	δ	θ_f
Face angle	Γ_0	δ_a
Pitch angle	Γ	δ'
Root angle	Γ_R	δ_t
Shaft angle	Σ	Σ

$$\sin \Gamma_2 = \frac{D_2}{2A_0} = \sin(\Sigma - \Gamma_3) = \sin \Sigma \cos \Gamma_3 - \cos \Sigma \sin \Gamma_3$$

$$\frac{1}{\tan \Gamma_2} = \frac{1}{\sin \Sigma}\left(\frac{\sin \Gamma_2}{\sin \Gamma_3} + \cos \Sigma\right)$$

$$\frac{D_2}{D_3} = \frac{\sin \Gamma_2}{\sin \Gamma_3}$$

Then

$$\tan \Gamma_2 = \frac{\sin \Sigma}{\cos \Sigma + D_2/D_3} = \frac{\sin \Sigma}{\cos \Sigma + N_2/N_3}$$

For manufacture of bevel gears the Gleason system has been adopted as a standard. For straight bevel gears the Gleason system tooth profiles are given in Table 10.8.

The teeth of a bevel gear do not have constant thickness along the tooth face. This is necessary since they are cut on the surface of a cone. In order to ensure that interference does not occur at the thinner sections of the tooth and also to provide for a larger radius on the edge of the cutter, the teeth are cut a little different from spur or helical teeth.

A section of two Gleason bevel gears is shown in Fig. 10.21. The dedendum elements are drawn toward the apex of the pitch cones, and addendum elements are drawn parallel to mating dedendum elements. The face width is limited as shown in Table 10.8 in order to avoid manufacturing difficulties.

10.15 SPECIAL BEVEL GEARS

Bevel gears assume various names depending upon their design or intended use.

Miter gears are gears of equal size operating with a shaft angle of 90°.

TABLE 10.8
Gleason straight bevel gear tooth proportions

	U.S. Standard	Metric
Addendum		
Gear	$A_G = \dfrac{0.540}{P} + \dfrac{0.460}{P(N_3/N_2)^2}$	$a_G = 0.540m + \dfrac{0.460m}{(N_3/N_2)^2}$
Pinion	$a_P = \dfrac{2.000}{P} - a_G$	$a_P = 2.000m - a_G$
Dedendum		
Gear	$b_G = \dfrac{2.188}{P} + 0.002 - a_G$	$b_G = 2.188m + 0.05 - a_G$
Pinion	$b_P = \dfrac{2.188}{P} + 0.002 - a_P$	$b_P = 2.188m + 0.05 - a_P$
Face width	$< 0.3A_0$	Smallest of $10.0/P$ or $10.0m$

Tooth thickness on pitch circle
Gear $\quad t_G = \dfrac{p}{2} - (a_P - a_G)\tan\phi$
Pinion $\quad t_P = p - t_G$
$p = $ circular pitch

Number of teeth	
Pinion	**Gear**
13	30 or more
14	20 or more
15	17 or more
16	16 or more

Angular bevel gears are gears of unequal size which generally operate with a shaft angle different from 90°.

Crown gears are bevel gears with pitch cone angles equal to 90°. The crown bevel gear is much like the spur gear rack.

Zerol bevel gears have curved teeth such that the spiral angle of the tooth is tangent to a cone element.

Spiral bevel gears have obliquely curved teeth. The spiral angle is such that the face advance of the tooth is greater than the circular pitch of the tooth. In this manner the spiral bevel gear is much like the helical gear.

Hypoid gears are spiral bevel gears which normally operate with 90° shaft angles. However, the shaft centers do not intersect. This form of bevel gear was introduced for use with automotive rear-axle drive systems known as the ring gear and pinion.

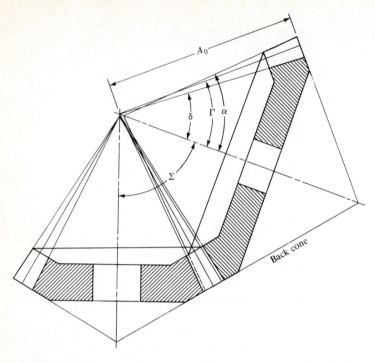

FIGURE 10.21
The Gleason system bevel gear.

PROBLEMS

10.1. For a 3-in pitch diameter U.S. standard gear with 60 teeth and a 20°
pressure angle, determine
(*a*) Diametral pitch
(*b*) Circular pitch
(*c*) Gear blank outside diameter
(*d*) Base circle radius

10.2. A British standard gear with 50 teeth is manufactured with a module of 6.
Determine
(*a*) Pitch diameter
(*b*) Base circle radius
(*c*) Circular pitch
(*d*) Outside diameter of gear blank

10.3. Two U.S. standard spur gears with $14\frac{1}{2}°$ pressure angle and 8 diametral
pitch are assembled in mesh with standard center distance. Gear 1 has 45
teeth and gear 2 has 63 teeth. Determine
(*a*) Pitch diameter of each gear
(*b*) Distance between centers of the gears

(c) Base circle radius of each gear

(d) Circular pitch of the gear teeth

(e) Outside diameter of each gear blank

(f) Angular velocity ratio of the assembly

10.4. Two British standard gears with module 8 are assembled in mesh with standard center distance. Gear 1 has 57 teeth and gear 2 has 128 teeth. Determine

(a) Pitch diameter of each gear

(b) Distance between centers of the gears

(c) Base circle radius of each gear

(d) Circular pitch of the gear teeth

(e) Outside diameter of each gear blank

(f) Angular velocity ratio of the assembly

10.5. A gear set consisting of U.S. standard gears with 20° pressure angle and 3 pitch is assembled with standard center distance. Gear 1 has 12 teeth and gear 2 has 24 teeth. Graphically lay out the pitch, base, and addendum circles, and

(a) Indicate the pitch point P

(b) Identify the points where contact begins and ends

(c) If interference exists, indicate the amount an addendum must be reduced to avoid interference

10.6. Two British standard gears are in mesh with standard center distance. The gears are cut with module 8. Gear 1 has 20 teeth and gear 2 has 32 teeth. Lay out the gears showing

(a) The pitch point P

(b) Pitch circles

(c) Base circles

(d) Line of action indicating start and end of contact

10.7. Two U.S. standard gears are cut with an 8-pitch 20° hob. Gear 1 has 20 teeth and gear 2 has 50 teeth. The gears are assembled with standard center distance. Calculate the gears' pitch and base circle radii, addendum, dedendum, circular pitch, and base circular pitch.

10.8. Two British standard gears are cut with 12 module. Gear 1 has 30 teeth and gear 2 has 55 teeth. The gears are assembled with standard center distance. Calculate the gears' pitch and base circle radii, addendum, dedendum, circular pitch, and base circular pitch.

10.9. Two U.S. standard gears are cut with a 12-pitch 20° hob. Gear 1 has 35 teeth, and gear 2 has 72 teeth. The gears are assembled with standard center distance. Determine the contact ratio.

10.10. Two British standard gears cut with 10 modules are assembled with standard center distance. Gear 1 has 47 teeth and gear 2 has 85 teeth. Calculate the contact ratio.

10.11. A U.S. standard spur gear with 20° pressure angle and 4 pitch has 38 teeth. Calculate the tooth thickness at a radius of 4.875 in.

10.12. A German standard metric gear has 56 teeth and 2.5 modules. Calculate the tooth thickness at a radius of 71.5 mm.

10.13. Two U.S. standard gears with 56 and 74 teeth, 12 pitch, and 20° pressure angle are assembled with standard center distance. If the center distance is increased 0.040 in, determine the resulting backlash.

10.14. Two parallel shafts are to be separated a distance of 9.25 in and coupled with gears providing an angular velocity ratio of 1.7. With a pitch of 10, recommend the numbers of teeth on each gear. If necessary recommend changes in the specifications.

10.15. A 37-tooth gear cut with a 10-pitch 20° hob has a helix angle of 21°. Calculate
  (a) Normal circular pitch
  (b) Circular pitch in the plane of rotation
  (c) Normal diametral pitch
  (d) Diametral pitch in the plane of rotation
  (e) Pitch diameter
  (f) Recommended face width
  (g) Outside diameter of gear blank

10.16. A 42-tooth helical gear with module 3 has a helix angle of 18°. Calculate
  (a) Normal circular pitch
  (b) Circular pitch in the plane of rotation
  (c) Normal module
  (d) Module in the plane of rotation
  (e) Pitch diameter
  (f) Recommended face width
  (g) Outside diameter of gear blank

10.17. Two parallel shafts are connected with helical gears which provide an angular velocity ratio of 3.6. Shaft center distance is 8.125, and the gears are cut with an 8-pitch hob. Calculate
  (a) Number of teeth on each gear
  (b) Helix angle
  (c) Outside diameter of each gear blank
  (d) Recommended face width

10.18. Two parallel shafts are to be separated a distance of 9.75 in and coupled with gears providing an angular velocity ratio of 1.7. The gears are to be cut with a 10-pitch hob. Recommend the gears including outside diameter of the gear blank and minimum face width.

10.19. Two shafts which intersect at an angle of 62° are to be connected by helical gears resulting in a 2:1 velocity ratio. The gears are cut with a 12-pitch hob. Gear 2 has a helix angle of 27° and a pitch diameter of 5.892 in its plane of rotation. Calculate the number of teeth on each gear and the pitch diameter and helix angle of gear 3 if the gears are of the same hand.

10.20. Shafts at 90° to each other are to be joined using helical gears giving an angular velocity ratio of 1.733:1. Center distance is 4.832 in. If the gears are to have equal helix angles and are cut with a 12-pitch hob, calculate the number of teeth on each gear.

10.21. A double thread worm has a pitch diameter of 2 in and an axial pitch of 0.625. Determine the lead angle.

10.22. A worm gear and worm with 90° shaft angle has a speed reduction of 20:1. The double thread worm has a lead angle of 15° and an axial pitch of 0.500. Determine
  (*a*) Pitch diameter of gear
  (*b*) Number of teeth on gear
  (*c*) Helix angle of gear
  (*d*) Center distance

10.23. A worm and worm gear are required with 90° shaft angle and 100:1 velocity ratio. Center distance between the shafts is to be 4.500 in. Design the worm and worm gear. (Some assumptions are necessary.)

10.24. A triple thread worm has a pitch diameter of 38 mm and an axial pitch of 16 mm. Determine the lead angle.

10.25. A worm and worm gear with 90° shaft angle is to have a double thread worm with a 15° lead angle and 10-mm pitch. Determine the required
  (*a*) Pitch diameter of the gear
  (*b*) Number of teeth on the gear
  (*c*) Helix angle of the gear
  (*d*) Center distance

10.26. A worm and worm gear are required with 90° shaft angle and an 80:1 speed reduction. Center distance is to be 300 mm. Design the worm and worm gear. (Some assumptions are necessary.)

10.27. A Gleason 8-pitch straight bevel pinion with 36 teeth drives a gear with 90 teeth. Shaft angle is 87°. Determine
  (*a*) Pitch cone angles
  (*b*) Addendums and dedendums
  (*c*) Pitch diameter
  (*d*) Face width of each gear

10.28. Two Gleason straight bevel gears are assembled with 90° shaft angle. The gears are module 10 with 42 and 79 teeth. Determine
  (*a*) Pitch cone angle
  (*b*) Addendums and dedendums
  (*c*) Recommended face width

CHAPTER
11

GEAR
TRAINS

11.1 INTRODUCTION

When two or more shafts are connected with gears, the resulting assembly is known as a *gear train*. Gear trains may contain spur gears, helical gears, bevel gears, or worms in various combinations. Gear trains are classified by type as shown in Fig. 11.1.

1. Simple gear trains are those in which each gear is mounted on a separate shaft.
2. Compound gear trains are those in which two or more gears share a common shaft.
3. Reverted gear trains are those in which two or more gear sets have a common center distance.
4. Planetary gear trains are those in which the axis of rotation of one or more gears is mounted on an arm which rotates about a center.

The sense of rotation of each gear is of importance. It is generally sufficient to recall that two gear teeth in mesh must move in the same direction. For many gear trains it is sufficient to assign a sign to CW rotation and recall that two gears in mesh rotate in opposite directions. Exceptions are encountered in use of internal, crossed helical, bevel, or worm gears. One technique which is particularly useful is to use arrows on a sketch of the gear train to indicate the direction of rotation as in Fig. 11.2.

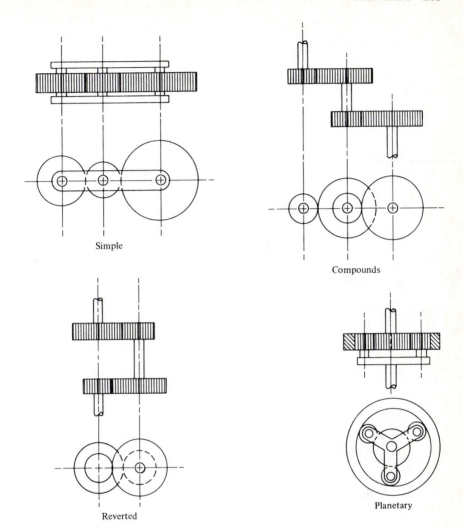

FIGURE 11.1
Gear train types.

11.2 SIMPLE GEAR TRAIN

In the simple gear train of Fig. 11.3, gear 2 is the input driver and it is desired to know the angular velocity and direction of rotation of gear 6. The angular velocity of gear 6 divided by the angular velocity of gear 2 is known as the *train value*. For a gear train it is only necessary to multiply the angular velocity of the input gear by the train value to find the angular velocity of the output gear.

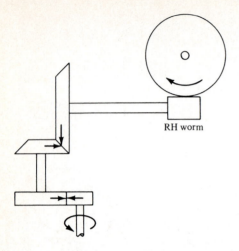

RH worm

FIGURE 11.2
Direction of rotation.

Two gears in mesh must have the same diametral pitch and pitch line velocity. Pitch diameter is related to the number of teeth on the gear by $N = PD$ or $D = N/P$. Then

$$V_p = \frac{D_2}{2}\,\omega_2 = \frac{D_3}{2}\,\omega_3$$

or

$$V_p = \frac{N_2}{2P}\,\omega_2 = \frac{N_3}{2P}\,\omega_3$$

where D_2, D_3 = pitch diameter of gears 2 and 3, respectively
N_2, N_3 = number of teeth on gears 2 and 3, respectively
P = pitch

and

$$\frac{\omega_2}{\omega_3} = \frac{N_3}{N_2}$$

Thus the angular velocity ratio between two gears in mesh is equal to the inverse ratio of the numbers of teeth on the gears.

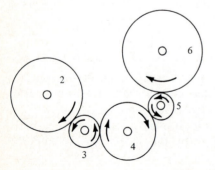

FIGURE 11.3
Angular velocity ratio.

Starting with the output gear and working toward the input gear with minus signs indicating that the mating gears rotate in opposite sense gives

$$\frac{\omega_6}{\omega_2} = \frac{\omega_6}{\omega_5}\frac{\omega_5}{\omega_4}\frac{\omega_4}{\omega_3}\frac{\omega_3}{\omega_2}$$

$$= \frac{-N_5}{N_6}\frac{-N_4}{N_5}\frac{-N_3}{N_4}\frac{-N_2}{N_3}$$

$$= \frac{+N_2}{N_6}$$

where the plus sign indicates that the output and input gears rotate in the same sense.

In the case of the simple gear train, the gears intermediate between the input and output gears are known as *idler gears* which simply serve to change the sense of rotation but do not influence the train value.

The train value of a gear train including a worm and worm gear may be found in a similar manner except the direction of rotation must be considered more carefully. In the system of Fig. 11.4, the worm is a double right-hand worm rotating in a CW sense when viewed from the motor end. The angular velocity of the output gear and its direction of

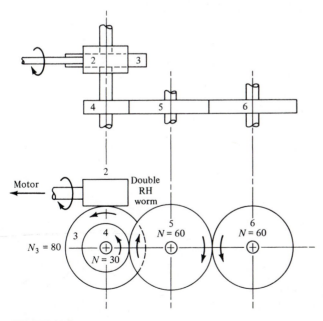

FIGURE 11.4
Gear train with a worm.

rotation are computed as follows

$$\text{Train value} = \text{TV} = \frac{\omega_6}{\omega_2} = \frac{N_2}{N_3} \frac{N_4}{N_5} \frac{N_5}{N_6}$$

$$= \frac{N_2}{N_3} \frac{N_4}{N_6}$$

$$= \frac{2}{80} \frac{30}{60}$$

$$\omega_6 = \omega_2 \frac{1}{80}$$

The direction of rotation is most easily found by using arrows on a sketch. Since the worm is a right-hand worm, if the worm wheel were held stationary and the worm rotated clockwise, the worm would tend to move to the right much like a cap screw being installed. Since the worm wheel is not held stationary and the worm cannot move to the right, the worm wheel gear teeth must move to the left or rotate counterclockwise. Gear 5 acts as an idler gear, and gear 6 will rotate in a CCW sense.

11.3 COMPOUND GEAR TRAIN

With two or more gears mounted firmly on the same shaft, each gear must have the same angular velocity and sense of rotation. In Fig. 11.5, gear 2 is the input, gears 3 and 4 share a common shaft known as a "jack shaft," and gear 5 is the output. The train value becomes ω_5/ω_2.

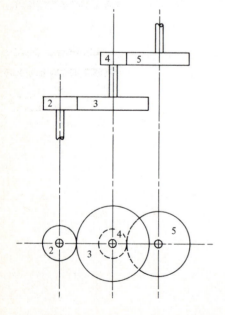

FIGURE 11.5
Compound gear train.

$$\frac{\omega_5}{\omega_2} = \frac{\omega_5}{\omega_4} \frac{\omega_4}{\omega_3} \frac{\omega_3}{\omega_2}$$

In this case, $\omega_4/\omega_3 = 1.0$ since both gears must rotate at the same angular velocity. Then

$$\frac{\omega_5}{\omega_2} = \frac{\omega_5}{\omega_4} \frac{\omega_3}{\omega_2}$$

$$= \frac{N_4}{N_5} \frac{N_2}{N_3}$$

The jack shaft acts as an idler gear and the output and input shafts rotate in the same sense.

It is interesting to note that gear 2 drives gear 3 and gear 4 drives gear 5. The train value for any compound gear train may be calculated from

$$TV = \frac{\text{product of number of teeth on driving gears}}{\text{product of number of teeth on driven gears}}$$

It is often desirable for the input and output shafts to share a common centerline. The reverted gear train of Fig. 11.6 is treated in exactly the same manner as that of Fig. 11.5. In this case it is necessary that the center distance between gears 2 and 3 be exactly the same as the center distance between gears 4 and 5. If the angular velocity ratio between gears 2 and 3 is different from that between gears 4 and 5, some complications may arise in the design of the system. Helical gears are generally used to ensure that the center distances are proper.

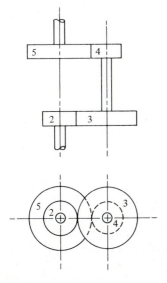

FIGURE 11.6
Reverted gear train.

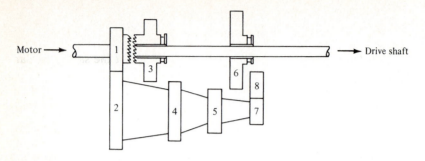

FIGURE 11.7
Three-speed automobile transmission.

A good example of the reverted gear train is the three-speed automobile transmission shown in Fig. 11.7. In first gear, gear 6 is moved to engage gear 5 and the power train is through gears 1, 2, 5, and 6 giving a train value of $N_1 N_5 / N_2 N_6$. In second gear, gear 3 is moved to engage gear 4 and the power train is through gears 1, 2, 4, and 3 giving a train value of $N_1 N_4 / N_2 N_3$. In third gear, gear 3 is moved to engage the teeth between gears 1 and 3 giving a train value of 1.0. In reverse gear, gear 6 is moved to engage idler gear 8 which is in contact with gear 7 giving a train value of $N_1 N_7 / N_2 N_6$.

11.4 PLANETARY SYSTEMS

If the center of rotation of one or more gears is allowed to rotate about another center, the resulting assembly is known as a *planetary* or *epicyclic system*. The gear whose center is not fixed but is free to move in a circular path about another center is known as a *planet gear*. For simplicity in sketching, only one planet gear is generally shown although two or more will be needed for dynamic balance. A gear about which planet gears move is known as a *sun gear*. The link carrying the planet is known as the *arm*.

Analysis of planetary systems is complicated by the fact that planet gears rotate about their own center while that center is moving about another center. Figure 11.8 shows a planet gear 2 attached to an arm 3 and able to move about the center of gear 4. If the assembly is welded together so that no relative motion can occur and is allowed to rotate one revolution about the center of gear 4, gear 2 will make one revolution about its own center while not rotating on its shaft.

The tabular method of analysis is illustrated in Fig. 11.9 where gear 4 provides the input angular rotation. Gear 2 is a planet gear (three gears all designated as gear 2 are shown for dynamic balance). The planet gears are mounted on arm 3 which is the output angular rotation of the system.

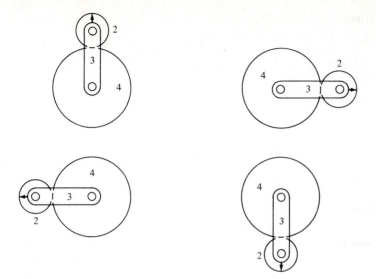

FIGURE 11.8
Planetary action.

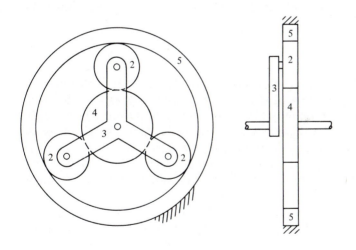

Steps	Gear			
	Arm (3)	2	4	5
Locked train	+1	+1	+1	+1
Fixed arm	0	$-N_5/N_2$	$+N_5/N_4$	-1
Total	+1	$1 - N_5/N_2$	$1 + N_5/N_4$	0

FIGURE 11.9
Planetary system.

Gear 5 is a fixed internal gear which does not rotate. In the tabular system a table is constructed showing the angular rotation of each gear and the arm. The entire assembly is considered to be locked to prevent any relative motion and is rotated one revolution CW or CCW. In this case it is essential to include a sign to show the sense of rotation.

The first step in tabular analysis is to consider the assembly locked and rotated one revolution. Each member of the system will make one revolution about its center of rotation. In the second step, the arm is considered to be fixed and all other gears are free to move. With the arm fixed, a simple gear train results. Since, in this example, gear 5 is fixed and not allowed to move, the gear 5 is rotated one revolution in a sense opposite to the rotation as a fixed assembly. In this manner, the sum of the first and second steps will be zero revolutions of gear 5. If the arm is fixed and gear 5 rotated one revolution, gear 2 will rotate N_5/N_2 revolutions and gear 4 will rotate $(N_5/N_2)(N_2/N_4)$ or N_5/N_4 revolutions. In this case gear 2 acts as an idler gear. Care should be taken to record the correct sense of rotation of each gear.

The sum of steps 1 and 2 is the final motion of each component of the system. Thus gear 5 will be fixed and not rotate. Gear 4 will rotate $1 + N_5/N_4$. Arm 3 will rotate 1.0. Gear 2 will rotate $1 - N_5/N_2$. The angular velocity of the output shaft divided by that of the input shaft will be

$$TV = \frac{1}{1 + N_5/N_4}$$

The tabular system is convenient and fast for use with relatively simple planetary systems having only one input. It is possible for a planetary system to have more than one input; that is, gear 5 need not be stationary but may be driven in a CW or CCW direction. If this is the case, it is necessary to use the tabular system twice: once with each drive system considered as stationary. The output of the two tabular systems may then be combined to give the total output with two inputs. It is also possible to conduct one tabulation using the actual revolutions of each gear.

The algebraic technique relies on fundamental principles of relative motion of three bodies.

1. If three bodies have relative angular velocities, the angular velocity of the first is equal to the angular velocity of the second relative to that of the first plus the angular velocity of the third relative to the second. If the bodies are numbered 2, 3, and 4,

$$\omega_{42} = \omega_{43} + \omega_{32}$$

2. If two bodies have relative angular velocities, the angular velocity of

the first with respect to the second is equal and opposite to the angular velocity of the second with respect to the first.

$$\omega_{32} = -\omega_{23}$$

With reference to Fig. 11.9, the ground is considered to be link 1. It is important to recognize the arm and consider rotation of the gears with respect to the arm. Thus,

$$\omega_{53} = \omega_{51} - \omega_{31}$$

and

$$\omega_{43} = \omega_{41} - \omega_{31}$$

Dividing the first equation by the second provides

$$\frac{\omega_{53}}{\omega_{43}} = \frac{\omega_{51} - \omega_{31}}{\omega_{41} - \omega_{31}}$$

In this equation ω_{53} and ω_{43} are both relative motions and ω_{51}, ω_{41}, ω_{31} are absolute motions since link 1 is the ground link. In evaluation of the equation ω_{53}/ω_{43} is the angular velocity ratio between links 5 and 4 if link 3 (the arm) is held stationary. If link 4 is considered to be the first gear in the train and link 5 is considered to be the last gear in the train with the arm lettered as A, the equation may be written as

$$\frac{\omega_{LA}}{\omega_{FA}} = \frac{\omega_L - \omega_A}{\omega_F - \omega_A}$$

where ω_{LA} = angular velocity of last gear with respect to arm
ω_{FA} = angular velocity of first gear with respect to arm
ω_L = angular velocity of last gear with respect to ground
ω_F = angular velocity of first gear with respect to ground
ω_A = angular velocity of the arm with respect to ground

In using this equation, the first and last gears must not be planet gears. The first and last gears must mesh with gears that have planetary motion. The first and last gears must also rotate about parallel axes. The direction of rotation must be properly designated by plus or minus signs.

In application of the equation to Fig. 10.9, the first gear is selected as gear 4 and the last gear is selected to be gear 5. The equation becomes

$$\frac{\omega_{53}}{\omega_{43}} = \frac{\omega_{51} - \omega_{31}}{\omega_{41} - \omega_{31}}$$

Then

$$\frac{\omega_{53}}{\omega_{43}} = \frac{\omega_5}{\omega_2}\left(-\frac{\omega_2}{\omega_4}\right) = -\frac{N_2}{N_5}\frac{N_4}{N_2} = -\frac{N_4}{N_5}$$

Since gear 5 is fixed and cannot move, $\omega_{51} = 0$ and

$$-\frac{N_4}{N_5} = \frac{0 - \omega_3}{\omega_4 - \omega_3}$$

Solving for ω_3/ω_4 gives

$$\frac{\omega_3}{\omega_4} = \frac{1}{1 + N_5/N_4}$$

11.5 EXAMPLES OF PLANETARY SYSTEMS

AUTOMOBILE DIFFERENTIAL. A schematic drawing of an automobile differential is shown in Fig. 11.10. Power from the engine is delivered through the drive shaft to the hypoid gear system gears 1 and 2. Gear 2 acts as the arm of a planetary system. The purpose of the differential is to allow one drive wheel to rotate at an angular velocity different from the other. This is necessary when a car is turning a corner since the outboard wheel must travel faster than the inner wheel to avoid scrubbing the tires.

When traveling in a straight line, there is no relative motion between gears 3 to 6. When turning, relative motion develops in the planetary system. Assume the angular velocity ratio between gears 1 and 2 is 4.0 and the drive shaft is turning 1600 r/min. With straight-line travel, the wheels will be revolving at 400 r/min. Assume then that during a turn the right wheel angular velocity becomes 350 r/min, and compute the angular velocity of the left wheel. Using the algebraic technique, gear 6 is designated as L, gear 4 is designated as F, and gear 2 is the arm. Then

$$\frac{\omega_{62}}{\omega_{42}} = \frac{\omega_6 - \omega_2}{\omega_4 - \omega_2} = -1$$

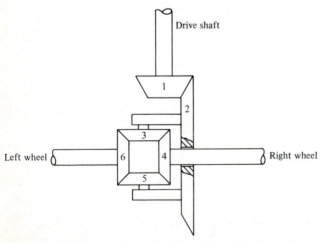

FIGURE 11.10
Differential system.

The ratio must be equal to -1 since gears 3 to 6 are all the same size and gear 3 simply functions as an idler gear. Substitution of known angular velocities gives

$$\frac{\omega_6 - 400}{350 - 400} = -1$$

and $\omega_6 = 450 \text{ r/min}$.

It is interesting to note that slowing of one wheel results in increasing the angular velocity of the other wheel an equal amount.

BEVEL GEAR SYSTEM. The use of bevel gears in planetary systems can result in very large speed reductions while using very little space. The Humpage system of Fig. 11.11 is an excellent example. Gear 2 is the driver of the system. Gear 6 is the output of the system. Arm 7 is free to rotate about the common centerline of gears 2 and 6. Gear 5 is fixed and cannot rotate. Analysis may be conducted using either the tabular technique or the algebraic technique.

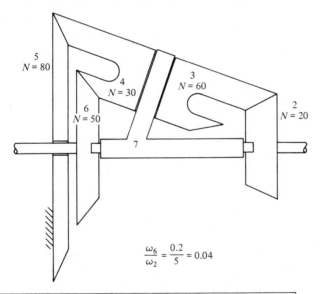

$$\frac{\omega_6}{\omega_2} = \frac{0.2}{5} = 0.04$$

Steps	2	3	4	5	6	Arm (7)
Locked train	+1	—	—	+1	+1	+1
Fixed arm	N_5/N_2	—	—	-1	$-N_5 N_4/N_3 N_6$	0
Total	5	—	—	0	0.2	1

FIGURE 11.11
Humpage planet system.

In using the tabular technique, a table is constructed in two steps as in Fig. 11.11.

The algebraic technique requires two steps. In the first step the train is considered to consist of gears 2, 3, and 5 and the angular motion of arm 7 is computed. Gear 2 is designated as F, and gear 5 is designated as L. Then

$$\frac{\omega_{27}}{\omega_{57}} = -\frac{N_5}{N_2} = -\frac{80}{20} = -4$$

and

$$-4 = \frac{\omega_2 - \omega_7}{\omega_5 - \omega_7}$$

For convenience consider the case in which ω_2 is 1000 r/min in a CW sense looking from the right. Remembering that $\omega_5 = 0$ gives

$$-4 = \frac{1000 - \omega_7}{0 - \omega_7}$$

or $\omega_7 = 200$ r/min CW (looking from the right).

In the second step the train is considered to consist of gears 2, 3, 4, 6, and arm 7. Gear 2 is designated as F and gear 6 is designated as L. Then

$$\frac{\omega_{27}}{\omega_{67}} = \frac{\omega_2 - \omega_7}{\omega_6 - \omega_7} = -\frac{N_3}{N_2}\frac{N_6}{N_4} = -5.0$$

Substituting $\omega_7 = 200$ and $\omega_2 = 1000$ gives $\omega_6 = 40$ r/min (same sense as gear 2).

COMPACT DRIVEHEAD SYSTEM. In order to insert rock bolts in the roof of mines it was necessary to devise a compact rock drill drive system. The system was required to rotate the drill and at the same time advance or remove the drill. The devised system is shown schematically in Fig. 11.12. The system consists of a simple gear train consisting of gears 1, 2, and 3 which function to rotate a hexagonal cross section drill bit. The simple system is driven by a reversible hydraulic motor shown as the drill motor. In order to advance or remove the drill bit, a second system consisting of gears 4 to 22 was superimposed on the simple gear train. The second system is driven by a reversible hydraulic motor designated as the differential motor.

Gears 14 to 22 are mounted on and rotate with gear 3. The angular velocity and direction of rotation of gears 21 and 22 is controlled by the direction of rotation and speed of the differential motor. A chain containing friction plates is connected between gears 21 and 22 and bear on the flat sides of the hexagonal drill bit. Thus the drill bit may be rotated using the drill motor and at the same time may be advanced or removed by using the differential motor.

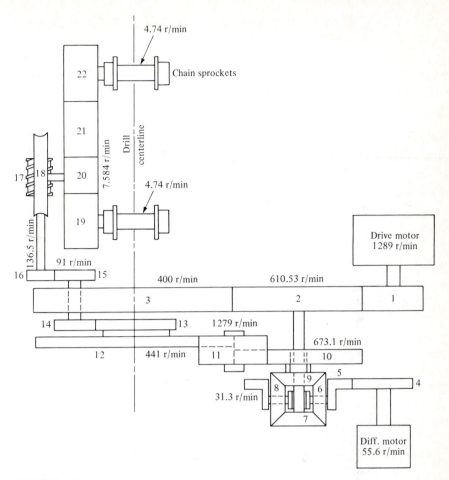

FIGURE 11.12
Drill drive system.

PROBLEMS

11.1. With an input of 100 r/min at gear 1, calculate the angular velocity and direction of rotation relative to gear 1 of gear 5. Numbers of teeth are shown in parentheses.

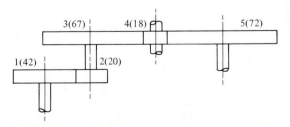

FIGURE P11.1
Simple gear train.

11.2. Gear 1 (worm) rotates in the direction shown at 100 r/min. Determine the direction of rotation and angular speed of gear 4.

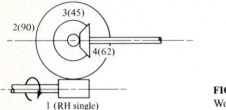

FIGURE P11.2
Worm gear system.

11.3. Input to the system is 100 r/min at gear 1 in the sense indicated. Determine the linear velocity and its direction for the rack (gear 7).

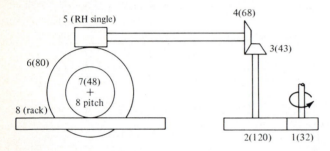

FIGURE P11.3
Worm and rack system.

11.4. Shafts *A* and *B* are connected with a jack shaft that contains two sliding gear clusters. Gear 1 can mesh with gear 2, or gear 4 can mesh with gear 3. Gear 5 can mesh with gear 6, or gear 8 can mesh with gear 7. Thus there are four possible output angular velocities for shaft *B*. Compute the output angular velocities possible with an input angular velocity of 100 r/min at shaft *A*.

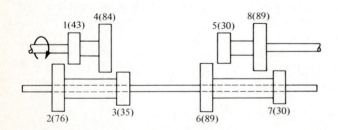

FIGURE P11.4
Transmission system.

11.5. Shaft A rotates at 1750 r/min. Determine the angular velocity of shaft B. Gear 4 is an internal gear and is fixed. Use the tabular technique.

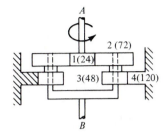

FIGURE P11.5
Planet system using tabular technique.

11.6. If the input to the planetary system through shaft A is 1800 r/min, determine the angular velocity of output shaft B and its sense of rotation relative to shaft A.

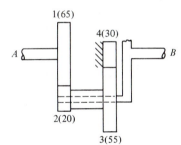

FIGURE P11.6
Planetary system.

11.7. In the automotive differential system, input is 750 r/min at the propeller shaft. If the left wheel is fixed and cannot turn, determine the angular velocity of the right wheel.

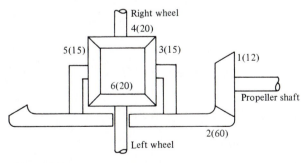

FIGURE P11.7
Differential system.

11.8. Input to the gear system is through shaft A at 1800 r/min. Determine the output angular velocity of shaft B.

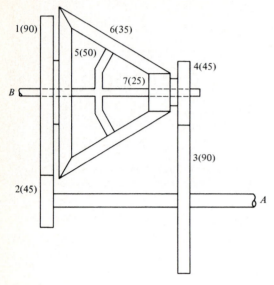

FIGURE P11.8
Planetary system with two inputs.

BIBLIOGRAPHY

A. HISTORY

W. H. G. Armytage, *A Social History of Engineering*, Faber & Faber, Ltd., London, 1961.

J. K. Finch, *The Story of Engineering*, Anchor Books, Doubleday & Company, Garden City, N.Y., 1960.

A. P. Usher, *A History of Mechanical Inventions*, revised edition, Harvard University Press, Cambridge, Mass., 1954.

Franz Reuleaux, *The Kinematics of Machinery*, Macmillan and Company, 1876. Reprinted, Dover Publications, New York, 1963.

R. Beyer, *Technische Kinematik*, Leipzig, 1931.

A. E. R. deJonge, "What Is Wrong with Kinematics and Mechanisms," *Mechanical Engineering*, vol. 64, no. 4, April 1942.

A. E. R. deJonge, "A Brief Account of Modern Kinematics," *ASME Transactions*, vol. 65, no. 6, August 1943.

R. S. Hartenberg and J. Denavit, *Kinematic Synthesis of Linkages*, McGraw-Hill Book Company, New York, 1964.

B. GRAPHICAL ANALYSIS

D. C. Tao, *Applied Linkage Synthesis*, Addison-Wesley Publishing Company, Reading, Mass., 1964.

G. H. Martin, *Kinematics and Dynamics of Machines*, McGraw-Hill Book Company, New York, 1969.

G. L. Guillet, *Kinematics of Machines*, John Wiley & Sons, New York, 1950.

J. Hirschhorn, *Kinematics and Dynamics of Plane Mechanisms*, McGraw-Hill Book Company, New York, 1962.

N. Rosenauer and A. H. Willis, *Kinematics of Mechanisms*, Associated General Publications Pty., Ltd., 1967. Available from Dover Publications, New York.

R. Beyer, *The Kinematic Synthesis of Mechanisms*. Translated from the German by H. Kuenzel, McGraw-Hill Book Company, New York, 1953.

K. Hain, *Applied Kinematics*, Translated from the German by D. P. Adams, F. R. E. Crossely, F. Freudenstein, T. P. Goodman, B. L. Harding, and D. R. Raichel, McGraw-Hill Book Company, New York, 1967.

C. ANALYTIC METHODS

A. H. Soni, *Mechanism Synthesis and Analysis*, Scripta Book Company, New York, 1974.

H. H. Mabie and C. F. Reinholts, *Mechanisms and Dynamics of Machinery*, John Wiley & Sons, New York, 1957.

C. H. Suh and C. W. Radcliffe, *Kinematics and Mechanisms Design*, John Wiley & Sons, New York, 1978.

G. N. Sandor and A. G. Erdman, *Advanced Mechanism Design*, vol. 2, Prentice-Hall, Englewood Cliffs, N.J., 1984.

M. L. James, G. M. Smith, and J. C. Wolford, *Applied Numerical Methods for Digital Computation*, IEP—A Dun-Donnelley Publisher, New York, 1977.

K. H. Hunt, *Kinematic Geometry of Mechanisms*, Oxford University Press, 1978.

D. CAMS

M. Kloomok and R. V. Muffley, "Plate Cam Design with Emphasis on Dynamic Effects," *Product Engineering*, February 1955.

M. Kloomok and R. V. Muffley, "Plate Cam Design," *Product Engineering*, September 1955. Revised by M. A. Ganter and J. J. Uicker, Jr., "Design Charts for Disk Cams with Reciprocating Radial Roller Followers," *ASME Transactions, Journal of Mechanical Design*, vol. 101, no. 3, July 1979.

F. Y. Chen, *Mechanics and Design of Cam Mechanisms*, Pergamon Press, New York, 1982.

H. A. Rothbart, *Cams*, John Wiley & Sons, New York, 1956.

D. Texar and G. K. Matthes, *The Dynamic Synthesis, Analysis, and Design of Modeled Cam Systems*, Lexington Books, Lexington, Mass., 1976.

E. GEARS

D. W. Dudley (ed.), *Gear Handbook*, McGraw-Hill Book Company, New York, 1962.

D. W. Dudley, *Practical Gear Design*, McGraw-Hill Book Company, New York, 1954.

AGMA Standards, New York.

APPENDIX
A

COMPUTER PROGRAMS

These programs are written in GW-BASIC language in a simplistic form which does not take advantage of such things as string functions and requires data input as separate items. With the programs in this form, students may use them directly or with only very minor modifications depending on the machine. GW-BASIC is a common language for many IBM-compatible personal computers.

The language requires a special program to evaluate the arccos of an angle as in lines 340 and 380 of the program ANGLES. Note that lines 290 to 320, 390, 400, 410, 440, 500, and 530 are devoted to providing the correct value of several angles and that all angles in all the programs are considered to be CCW positive from the x axis through the fixed link. Lines 390, 410, 420, 480, 500 and 510 of the program ANGLES provide integer values of the angles. If more significant figures are desired, the word INT must be removed from these lines in the program. Remember that each PRINT line must be preceded with L in order for the program output to print out to a printer.

```
10    PRINT "****************** ANGLES *********************"
20    PRINT "THIS PROGRAM PROVIDES ANGLES WITH THE HORIZONTAL"
30    PRINT "(COUNTERCLOCKWISE POSITIVE) FOR EACH LINK OF AN"
40    PRINT "OPEN OR CROSSED MECHANISM. ALL ANGLES ARE IN DEGREES"
50    PRINT
60    INPUT " LENGTH OF FIXED LINK =";P
70    INPUT " LENGTH OF CRANK =";Q
80    INPUT " LENGTH OF COUPLER =";R
90    INPUT " LENGTH OF ROCKER =";S
100   INPUT " CRANK ANGLE STEPS=";E
110   INPUT "OPEN(1) OR CROSSED (2) MECHANISM";Z
120   PRINT "LENGTH OF FIXED LINK =";P
130   PRINT "LENGTH OF CRANK       =";Q
140   PRINT "LENGTH OF COUPLER     =";R
150   PRINT "LENGTH OF ROCKER      =";S
160   PRINT
170   IF Z=1 THEN PRINT "MECHANISM IS OPEN"
180   IF Z=2 THEN PRINT "MECHANISM IS CROSSED"
190   PRINT
200   PRINT "CRANK          COUPLER         ROCKER         TRANSMISSION"
210   PRINT "ANGLE          ANGLE           ANGLE              ANGLE"
220   PRINT
230   FOR T=0 TO 360 STEP E
240   W=T*.0174532
250   D2=P*P + Q*Q - 2*P*Q*COS(W)
260   D=SQR(D2)
270   SIGY = -Q*SIN(W)
280   SIGX = P - Q*COS(W)
290   SIGMA=ABS(ATN(SIGY/SIGX))
300   IF SIGY<0 THEN IF SIGX<0 THEN SIGMA=SIGMA + 3.1415927
310   IF SIGY>0 THEN IF SIGX<0 THEN SIGMA=ABS(SIGMA+1.5707963)
320   IF SIGY<0 THEN IF SIGX>0 THEN SIGMA=6.2831853 - SIGMA
330   GAM=(D2 + R*R - S*S)/(2*D*R)
340   GAMMA=-ATN(GAM/SQR(-GAM*GAM + 1)) + 1.5707963
350   IF Z=2 GOTO 460
360   EPSILON=GAMMA + SIGMA - 6.2831853
370   BET=(R*R + S*S - D2)/(2*D*R)
380   BETA=-ATN(BET/SQR(-BET*BET + 1)) + 1.5707963
390   ALPHA=3.1415927 - GAMMA - BETA
400   PHI = INT((SIGMA + GAMMA + BETA -6.2831853)/.0174532)
410   IF PHI<0 THEN PHI = 360 + PHI
420   BETA = INT(BETA/.0174532)
```

```
430    EPSILON=INT(EPSILON/.0174532)
440    IF EPSILON<0 THEN EPSILON = 360 + EPSILON
450    GOTO 540
460    EPSILON = SIGMA - GAMMA
470    BET=(R*R+ S*S - D2)/(2*R*S)
480    BETA= -ATN(BET/SQR(-BET*BET + 1)) + 1.5707963
490    PHI=((SIGMA - GAMMA - BETA)/.0174532)
500    IF PHI<0 THEN PHI = 360 + PHI
510    BETA= INT(BETA/.0174532)
520    EPSILON = INT(EPSILON/.0174532)
530    IF EPSILON<0 THEN EPSILON = 360 + EPSILON
540    PRINT T,EPSILON, PHI, BETA
550    NEXT T
560    END
```

The following additions and modifications should be made to the ANGLES program to give the coordinates of a coupler point as a function of the input crank angle.

```
10     PRINT "***************** COUPLER *********************"

105    INPUT "DISTANCE ALONG COUPLER TO PERPENDICULAR TO COUPLER
       POINT";R1
106    INPUT "PERPENDICULAR DISTANCE FROM COUPLER TO COUPLER POINT"
       ;R2
155    PRINT "DISTANCE ALONG COUPLER TO PERPENDICULAR TO COUPLER
       POINT";R1
156    PRINT "PERPENDICULAR DISTANCE FROM COUPLER TO COUPLER POINT"
       ;R2
435    CX=Q*COS(W)+R1*COS(EPSILON*.0174532)-R2*SIN((EPSILON+90)*
       .0174532)
436    CY=Q*SIN(W)+R1*SIN(EPSILON*.0174532)+R2*COS((EPSILON+90)*
       .0174532)
525    CX=Q*COS(W)+R1*COS(EPSILON*.0174532)-R2*SIN((EPSILON+90)*
       .0174532)
526    CY=Q*SIN(W)+R1*SIN(EPSILON*.0174532)+R2*COS((EPSILON+90)*
       .0174532)
540    PRINT T,EPSILON,PHI,BETA,CX,CY
```

The program VELOCITY gives the angular velocity of each link of a four-bar mechanism.

```
10    PRINT "*************** VELOCITY ****************"
20    PRINT "THIS PROGRAM PROVIDES ANGULAR VELOCITY OF EACH"
30    PRINT "LINK (COUNTERCLOCKWISE POSITIVE) OF AN OPEN OR"
40    PRINT "CROSSED 4-BAR MECHANISM. CRANK ANGLE IN DEGREES"
50    PRINT "VELOCITIES IN RADIANS PER SECOND"
60    PRINT
70    INPUT "LENGTH OF FIXED LINK = ";P
80    INPUT "LENGTH OF CRANK =";Q
90    INPUT "LENGTH OF COUPLER =";R
100   INPUT "LENGTH OF ROCKER =";S
110   INPUT "CRANK ANGULAR VELOCITY =";OMQ
120   INPUT "CRANK ANGLE STEPS =";E
130   INPUT "OPEN (1) OR CROSSED (2) MECHANISM";Z
140   PRINT "LENGTH OF FIXED LINK =";P
150   PRINT "LENGTH OF CRANK      =";Q
160   PRINT "LENGTH OF COUPLER    =";R
170   PRINT "LENGTH OF ROCKER     =";S
180   PRINT "CRANK ANGULAR VELOCITY =";OMQ
190   PRINT
200   IF Z=1 THEN PRINT "MECHANISM IS OPEN" ELSE PRINT
      "MECHANISM IS CROSSED"
210   PRINT
220   PRINT "CRANK           COUPLER        ROCKER"
230   PRINT "ANGLE           ANGULAR        ANGULAR"
240   PRINT "                VELOCITY       VELOCITY"
250   PRINT
260   FOR T = 0 TO 360 STEP E
270   W=T*.0174532
280   D2 = P*P + Q*Q - 2*P*Q*COS(W)
290   D = SQR(D2)
300   SIGY = -Q*SIN(W)
310   SIGX = P - Q*COS(W)
320   SIGMA = ABS(ATN(SIGY/SIGX))
330   IF SIGY<0 THEN IF SIGX<0 THEN SIGMA=SIGMA+3.1415927
340   IF SIGY>0 THEN IF SIGX<0 THEN SIGMA=SIGMA+1.5707963
350   IF SIGY<0 THEN IF SIGX>0 THEN SIGMA=6.2831853 - SIGMA
360   GAM=(D2 + R*R - S*S)/(2*D*R)
370   GAMMA=-ATN(GAM/SQR(-GAM*GAM+1))+1.5707963
380   BET=(R*R + S*S - D2)/)2*R*S)
390   BETA = -ATN(BET/SQR(-BET*BET + 1)) + 1.5707963
400   IF Z=2 GOTO 490
410   EPSILON=GAMMA + SIGMA - 6.2831853
```

```
420   ALPHA=3.1415927 - GAMMA - BETA
430   PHI=SIGMA + GAMMA + BETA - 6.2831853
440   IF PHI<0 THEN PHI= 6.2831853 + PHI
450   IF EPSILON<0 THEN EPSILON= 6.2831853 + EPSILON
460   OMR=(OMQ*Q*SIN(W-PHI))/(R*SIN(PHI-EPSILON))
470   OMS=(OMQ*Q*SIN(W-EPSILON))/(S*SIN(PHI-EPSILON))
480   GOTO 550
490   EPSILON=SIGMA-GAMMA
500   PHI=SIGMA - GAMMA - BETA
510   IF PHI<0 THEN PHI = 6.2831853 + PHI
520   IF EPSILON<0 THEN EPSILON = 6.2831853 + EPSILON
530   OMR=(OMQ*Q*SIN(W-PHI))/(R*SIN(PHI-EPSILON))
540   OMS=(OMQ*Q*SIN(W-EPSILON))/(S*SIN(PHI-EPSILON))
550   PRINT T,OMR,OMS
560   NEXT T
570   END
```

The program VELOCITY may be modified with the following statements to return the velocity of a coupler point.

```
101   INPUT "DISTANCE ALONG COUPLER FROM CRANK TO
      PERPENDICULAR TO COUPLER POINT";R1
102   INPUT "PERPENDICULAR DISTANCE FROM COUPLER TO COUPLER
      POINT";R2
171   PRINT "DISTANCE ALONG COUPLER FROM CRANK TO
      PERPENDICULAR TO COUPLER POINT =";R1
172   PRINT "PERPENDICULAR DISTANCE FROM COUPLER TO COUPLER
      POINT =";R2
230   PRINT "CRANK       COUPLER      ROCKER      COUPLER POINT"
240   PRINT "ANGLE       ANGULAR      ANGULAR       VELOCITY"
250   PRINT "            VELOCITY    VELOCITY  VX    VY    V"
391   RC=SQR(R1*R1 + R2*R2)
392   RX=R1*COS(EPSILON) + R2*COS(EPSILON + 6.2831853)
393   RY=R1*SIN(EPSILON) + R2*SIN(EPSILON + 6.2831853)
394   NU=ATN(RY/RX)
395   IF R1>0 THEN IF R2<0 THEN NU=NU + 4.712389
396   IF R1<0 THEN IF R2>0 THEN NU=NU + 6.2831853
397   IF R1>0 THEN IF R2<0 THEN NU=NU + 3.1415927
471   VCX=-Q*OMQ*SIN(W)-RC*OMR*SIN(NU)
472   VCY=Q*OMQ*COS(W)+RC*OMR*COS(NU)
473   VC=SQR(VCX^2 + VCY^2)
541   VCX=-Q*OMQ*SIN(W)-RC*OMR*SIN(NU)
```

```
542  VCY=Q*OMQ*COS(W)+RC*OMR*COS(NU)
543  VC=SQR(VCX^2 + VCY^2)
550  PRINT T,OMR,OMS,VCX,VCY,VC
```

This program will give the angular acceleration of each link of a four-bar mechanism.

```
10 PRINT "THIS PROGRAM PROVIDES ANGULAR ACCELERATION OF EACH"
20 PRINT "LINK OF A 4-BAR MECHANISM (COUNTERCLOCKWISE POSITIVE)"
30 PRINT "OPEN OR CROSSED MECHANISMS ARE CONSIDERED.  VELOCITIES"
40 PRINT "ARE IN RADIANS/SEC.  ACCELERATIONS IN RADIANS/SEC. SQ."
50 PRINT
60 INPUT "LENGTH OF FIXED LINK";P
70 INPUT "LENGTH OF CRANK";Q
80 INPUT "LENGTH OF COUPLER";R
90 INPUT "LENGTH OF ROCKER";S
100 INPUT "CRANK ANGULAR VELOCITY, RADIANS/SECOND";OMQ
110 INPUT "CRANK ANGULAR ACCELERATION, RADIANS/SEC SQUARE";ALQ
120 INPUT "OPEN (1) OR CROSSED (2) MECHANISM";Z
130 INPUT "CRANK ANGLE STEPS";E
140 PRINT "LENGTH OF FIXED LINK =";P
150 PRINT "LENGTH OF CRANK        =";Q
160 PRINT "LENGTH OF COUPLER     =";R
170 PRINT "LENGTH OF ROCKER      =";S
180 PRINT
190 PRINT "CRANK ANGULAR VELOCITY      =";OMQ
200 PRINT "CRANK ANGULAR ACCELERATION =";ALQ
210 PRINT
220 IF  Z=1 THEN PRINT "MECHANISM IS OPEN" ELSE PRINT "MECHANISM
    IS CROSSED"
230 PRINT
240 PRINT "CRANK           COUPLER              ROCKER"
250 PRINT "ANGLE           ANGULAR              ANGULAR"
260 PRINT "                ACCELERATION     ACCELERATION"
270 PRINT
280 FOR T = 0 TO 360 STEP E
290 W=T*.0174532
300 D2 = P*P + Q*Q - 2*P*Q*COS(W)
310 D = SQR(D2)
320 SIGY = -Q*SIN(W)
330 SIGX = P - Q*COS(W)
340 SIGMA = ABS(ATN(SIGY/SIGX))
```

```
350 IF SIGY<0 THEN IF SIGX<0 THEN SIGMA = SIGMA + 3.1415927
360 IF SIGY>0 THEN IF SIGX<0 THEN SIGMA = SIGMA + 1.5707963
370 IF SIGY<0 THEN IF SIGX>0 THEN SIGMA = 6.2831853 - SIGMA
380 GAM=(D2 + R*R - S*S)/(2*D*R)
390 GAMMA = -ATN(GAM/SQR(-GAM*GAM +1)) + 1.5707963
400 BET = (R*R + S*S - D2)/)2*R*S)
410 BETA = - ATN(BET/SQR(-BET*BET + 1)) + 1.5707963
420 IF Z=2 GOTO 560
430 EPSILON = GAMMA + SIGMA - 6.2831853
440 ALPHA = 3.1415927 - GAMMA - BETA
450 PHI = SIGMA + GAMMA + BETA - 6.2831853
460 IF PHI<0  THEN PHI = 6.2831853 + PHI
470 IF EPSILON<0 THEN EPSILON = 6.2831853 + EPSILON
480 OMR = (Q*OMQ*SIN(W-PHI))/(R*SIN(PHI-EPSILON))
490 OMS = (Q*OMQ*SIN(W-EPSILON))/(S*SIN(PHI-EPSILON)
500 C = Q*OMQ*OMQ*SIN(W) - Q*ALQ*COS(W) + R*OMR*OMR*SIN(EPSILON)
    -S*OMS*OMS*SIN(PHI)
510 H = S*OMS*OMS*COS(PHI) - Q*ALQ*SIN(PHI) - Q*OMQ*OMQ*COS(W)
    -R*OMR*OMR*COS(EPSILON)
520 DEN = R*S*SIN(EPSILON - PHI)
530 ALR= (-S*C*SIN(PHI) + H*S*COS(PHI))/DEN
540 ALS= (H*R*COS(EPSILON) - C*R*SIN(EPSILON))/DEN
550 GOTO 670
560 EPSILION = SIGMA - GAMMA
570 PHI = SIGMA - GAMMA - BETA
580 IF PHI<0 THEN PHI = 6.2831853 + PHI
590 IF EPSILON<0 THEN EPSILON = 6.2831853 + EPSILON
600 OMR = (Q*OMQ*SIN(W-PHI))/(R*SIN(PHI-EPSILON))
610 OMS = (Q*OMQ*SIN(W-EPSILON))/(S*SIN(PHI-EPSILON))
620 C = Q*OMQ*OMQ*SIN(W)-Q*ALQ*COS(W)+R*OMR*OMR*SIN(EPSILON)
    -S*OMS*OMS*SIN(PHI)
630 H = S*OMS*OMS*COS(PHI)-Q*ALQ*SIN(PHI)-Q*OMQ*OMQ*COS(W)
    -R*OMR*OMR*COS(EPSILON)
640 DEN = R*S*SIN(EPSILON-PHI)
650 ALR = (-S*C*SIN(PHI) + H*S*COS(PHI))/DEN
660 ALS = (H*R*COS(EPSILON) - C*R*SIN(EPSILON))/DEN
670 PRINT T,ALR,ALS
680 NEXT T
690 END
```

The program for angular acceleration of links may be modified as follows to provide the acceleration of a coupler point.

```
10    PRINT "*********** COUPLER POINT ACCELERATION ************"
20    PRINT "THIS PROGRAM PROVIDES THE ACCELERATION OF A COUPLER"
30    PRINT "POINT OF AN OPEN OR CROSSED 4-BAR MECHANISM"
91    INPUT "DISTANCE FROM CRANK PIN TO PERPENDICULAR TO COUPLER
           POINT";R1
92    INPUT "PERPENDICULAR DISTANCE FROM COUPLER TO COUPLER
           POINT";R2
181   PRINT "DISTANCE FROM CRANK PIN TO PERPENDICULAR TO COOUPLER
           POINT";R1
182   PRINT "PERPENDICULAR DISTANCE FROM COUPLER TO COUPLER
           POINT";R2
240   PRINT "CRANK       C O U P L E R    P O I N T"
250   PRINT "ANGLE          X  ACCEL    Y ACCEL"
260   PRINT
270   RC=SQR(R1*R1 + R2*R2)
271   NU=ATN(R2/R1)
272   IF R1<0 THEN IF R2>0 THEN NU= 3.1415927 + NU
273   IF R1<0 THEN IF R2<0 THEN NU= 3.1415927 + NU
274   IF R1>0 THEN IF R2<0 THEN NU= 6.2831853 + NU
541   ACX=-Q*ALQ*SIN(W)-Q*OMQ*OMQ*COS(W)-RC*ALR*SIN(EPSILON+NU)
          -RC*OMR*OMR*COS(EPSILON+NU)
542   ACY=Q*ALQ*COS(W)-Q*OMQ*OMQ*SIN(W)+RC*ALR*COS(EPSILON+NU)
          -RC*OMR*OMR*SIN(EPSILON+NU)
661   ACX=-Q*ALQ*SIN(W)-Q*OMQ*OMQ*COS(W)-RC*ALR*SIN(EPSILON+NU)
          -RC*Q*ALQ*COS(W)-Q*OMQ*OMQ*SIN(W)+RC*ALR*COS(EPSILON+NU)
670   PRINT T,ACX, ACY
```

If the following statements are added to the COUPLER POINT ACCELERATION program, it will return the acceleration of the center of gravity of each link of a four-bar mechanism.

```
INPUT "DISTANCE FROM INPUT CRANK PIVOT TO CENTER OF GRAVITY
       OF INPUT CRANK";RG2
INPUT "DISTANCE FROM OUTPUT ROCKER PIVOT TO CENTER OF
       GRAVITY OF OUTPUT ROCKER";RG4
AG2X = -RG2*OMQ*OMQ*COS(W)-RG2*ALQ*SIN(W)
AG2Y = -RG2*OMQ*OMQ*SIN(W)+RG2*ALQ*COS(W)
AG4X = -RG4*OMS*OMS*COS(PHI)-RG4*ALS*SIN(PHI)
AG4Y = -RG4*OMS*OMS*SIN(PHI)+RG4*ALS*COS(PHI)
```

The print statement should be modified to print AG2X, AG2Y, AG4X, AG4Y in addition to acceleration of the coupler link center of gravity.

The program "LOOPS" is used to synthesize a four-bar mechanism for three positions of the mechanism with the second and third positions measured from the first position of the mechanism.

```
10   PRINT "LOOPS"
20   PRINT "THIS PROGRAM SOLVES EQUATIONS 21,22,23,24"
30   PRINT
40   K=0.01745
50   INPUT "THETA 12 =";T12
60   T12=T12*K
70   INPUT "THETA 13 =";T13
80   T13=T13*K
90   INPUT "THETA 32 =";T32
100  T32=T32*K
110  INPUT "THETA 33 =";T33
120  T33=T33*K
130  INPUT "THETA 52 =";T52
140  T52=T52*K
150  INPUT "THETA 53 =";T53
160  T53=T53*K
170  INPUT "DELTA 2 =";DEL2
180  INPUT "ARGUMENT OF DELTA 2 =";AL2
190  AL2=AL2*K
200  INPUT "DELTA 3 =";DEL3
210  INPUT "ARGUMENT OF DELTA 3 =";AL3
220  AL3=AL3*K
230  D2R=COS(T12+T53)-COS(T13+T52)+COS(T13)+COS(T52)-COS(T12)
       -COS(T53)
240  D2I=SIN(T12+T53)-SIN(T13+T52)+SIN(T13)+SIN(T52)-SIN(T12)
       - SIN(T53)
250  D2T=D2R*D2R+D2I*D2I
260  R1R=DEL2*(COS(AL2+T53)-COS(AL2))-DEL3*(COS(AL3+T52)-COS(AL3))
270  R1I=DEL2*(SIN(AL2+T53)-SIN(AL2))-DEL3*(SIN(AL3+T52)-SIN(AL3))
280  R1RE=(R1R*D2R+R1I*D2I)/D2T
290  R1IM=(R1I*D2R-R1R*D2I)/D2T
300  R5R=DEL3*(COS(AL3+T12)-COS(AL3))-DEL2*(COS(AL2+T13)-COS(AL2))
310  R5I=DEL3*(SIN(AL3+T12)-SIN(AL3))-DEL2*(SIN(AL2+T13)-SIN(AL2))
320  R5RE=(R5R*D2R+R5I*D1I)/D2T
330  R5IM=(R5I*D2R-R5R*D2I)/D2T
340  D3R=COS(T32+T53)-COS(T33+T52)+COS(T33)+COS(T52)-COS(T32)
       -COS(T53)
350  D3I=SIN(T32+T53)-SIN(T33+T52)+SIN(T33)+SIN(T52)-SIN(T32)
       -SIN(T53)
```

```
360 D3T=D3R*D3R+D3I*D3I
370 R3RE=(R3R*D3R+R3I*D3I)/D3T
380 R3IM=(REI*D3R-R3R*DEI)/D3T
390 R6R=DEL3*(COS(AL3+T32)-COS(AL3))-DEL2*(COS(AL2+T33)-COS(AL2))
400 R6I=DEL3*(SIN(AL3+T32)-SIN(AL3))-DEL2*(SIN(AL2+T33)-SIN(AL2))
410 R6RE=(R6R*D3R+R6I*D3I)/D3T
420 R6IM=(R6I*D3R-R6R*D3I)/D3T
430 R2RE=R5RE-R6RE
440 R2IM=R5IM-R6IM
450 R4RE=R1RE+R2RE-R3RE
460 R4IM=R1IM+R2IM-R3IM
470 PRINT"LINK NO.    LINK COMPONENT"
480 PRINT"            REAL     IMAGINARY     TOTAL"
490 PRINT"  1   ",  R1RE,  R1IM,  SQR(R1RE*R1RE+R1IM*R1IM)
500 PRINT"  2   ",  R2RE,  R2IM,  SQR(R2RE*R2RE+R2IM*R2IM)
510 PRINT"  3   ",  R3RE,  R3IM,  SQR(R3RE*R3RE+R3IM*R3IM)
520 PRINT"  4   ",  R4RE,  R4IM,  SQR(R4RE*R4RE+R4IM*R4IM)
530 PRINT"  5   ",  R5RE,  R5IM,  SQR(R5RE*R5RE+R5IM*R5IM)
540 PRINT"  6   ",  R6RE,  R6IM,  SQR(R6RE*R6RE+R6IM*R6IM)
550 PRINT
560 INPUT"DO YOU WANT TO DO ANOTHER MECHANISM Y OR N";P
570 IF P=Y THEN 10 ELSE 580
580 END
```

The program NONLINEAR uses the Newton-Rapson method to find the angular relationship described by the Freudenstein equation.

```
10     PRINT "*************** NONLINEAR *******************"
20     PRINT "THIS PROGRAM PROVIDES THE ROCKER ANGLE AS A FUNCTION
           OF THE"
30     PRINT "CRANK ANGLE FOR A 4-BAR MECHANISM USING THE
           FREUDENSTINE"
40     PRINT "EQUATION AND THE NEWTON-RAPSON METHOD"
50     PRINT
60     INPUT "CRANK LENGTH";L1
70     INPUT "COUPLER LENGTH";L2
80     INPUT "ROCKER LENGTH";L3
90     INPUT "FIXED LINK LENGTH";L4
100    INPUT "ALLOWABLE ERROR IN RADIANS";E
110    PRINT "CRANK LENGTH        =";L1
120    PRINT "COUPLER LENGTH      =";L2
130    PRINT "ROCKER LENGTH       =";L3
```

```
140    PRINT "FIXED LINK LENGTH    =";L4
150    PRINT
160    INPUT "INITIAL ESTIMATE OF THETA 3";T3
170    INPUT "STEPS OF CRANK ANGLE";S
180    PRINT "CRANK        ROCKER"
190    PRINT "ANGLE        ANGLE"
200    PRINT
210    R1=(L4*L4 + L3*L3 + L1*L1 - L2*L2)/(2*L1*L3)
220    R2=L4/L1
230    R3=L4/L3
240    FOR T1=0 TO 360 STEP S
250    F=R1+R2*COS(T3*.01745)-R3*COS(T1*.01745)
         -COS(T1*.01745-T3*.01745)
260    DF=R2*SIN(T3*.01745)-SIN(T1*.01745-T3*.01745)
270    T31=T3-T/DF
280    E1=ABS(T31-T3)
290    IF E1<E THEN GOTO 320 ELSE GOTO 300
300    T3=T31
310    GOTO 250
320    T3=T31
330    PRINT T1,T3
340    NEXT T1
350    END
```

The CAMPA program gives the pressure angle for each cam angle for a roller follower cam with cycloidal or simple harmonic motion.

```
10   PRINT "CAMPA PROGRAM"
20   PRINT "ROLLER FOLLOWER CAM PRESSURE ANGLE WITH CYCLOIDAL"
30   PRINT "OR SIMPLE HARMONIC MOTION"
40   PRINT
50   INPUT "CAM ANGLE FOR TOTAL MOTION, DEGREES";B
60   INPUT "FOLLOWER LIFT";L
70   INPUT "PITCH BASE CIRCLE RADIUS";RB
80   INPUT "FOLLOWER OFFSET";E
90   INPUT "CYCLOIDAL (1) OR HARMONIC (2)";K
100  PRINT
110  IF K=1 THEN PRINT "CYCLOIDAL FOLLOWER MOTION"
120  IF K=2 THEN PRINT "SIMPLE HARMONIC FOLLOWER MOTION"
130  PRINT "CAM ANGLE FOR TOTAL MOTION, DEGREES";B
140  PRINT "FOLLOWER LIFT               ";L
150  PRINT "PITCH BASE CIRCLE RADIUS        ";RB
```

```
160 PRINT "FOLLOWER OFFSET                    ";E
170 PRINT
180 PRINT "CAM ANGLE     PRESSURE ANGLE"
190 IF K=1 THEN GOTO 280
200 FOR T=0 TO B STEP 10
210 S=(.5*L)*(1-COS(3.14159*T/B))
220 DS=(1.5758*L*SIN(3.14159*T/B))/(B*.01745)
230 R=SQR(E^2+(RB+S)^2)
240 ALPHA=ATN(DS/R)
250 PRINT T, ALPHA/.01745
260 NEXT T
270 END
280 FOR T=0 TO B STEP 10
290 S=L*((T/B)-(.1592*SIN(6.2832*T/B)))
300 DS=(L/(B*.01745))*(1-COS(6.2832*T/B))
310 R=SQR(E^2+(RB+S)^2)
320 ALPHA = ATN(DS/R)
330 PRINT T,ALPHA/.01745
340 NEXT T
350 END
```

The program SIMULT will solve up to nine simultaneous linear equations.

```
10    PRINT "**************** SIMULT *****************"
20    PRINT
30    REM - LIMIT A() TO A(R,R+1) WHERE R=MAX. NO. OF EQUATIONS
40    DIM A(9,10)
50    INPUT "NUMBER OF EQUATIONS";R
60    PRINT "COEFFICIENT MATRIX"
70    FOR J=1 TO R
80    PRINT "EQUATION";J
90    FOR I=1 TO R+1
100   IF I=R+1 THEN 130
110   PRINT "COEFFICIENT";I;
120   GOTO 140
130   PRINT "CONSTANT";
140   INPUT A(J,I)
150   NEXT I
160   NEXT J
170   FOR J=1 TO R
180   FOR I=J TO R
```

```
190   IF A(I,J)<>0 THEN 230
200   NEXT I
210   PRINT "NO UNIQUE SOLUTION"
220   GOTO 440
230   FOR K=1 TO R+1
240   X=A(J,K)
250   A(J,K)=A(I,K)
260   A(I,K)=X
270   NEXT K
280   Y=1/A(J,J)
290   FOR K=1 TO R+1
300   A(J,K)=Y*A(J,K)
310   NEXT K
320   FOR I=1 TO R
330   IF I=J THEN 380
340   Y=-A(I,J)
350   FOR K=1 TO R+1
360   A(I,K)=A(I,K)+Y*A(J,K)
370   NEXT K
380   NEXT I
390   NEXT J
400   PRINT
410   FOR I=1 TO R
420   PRINT "X";I;"=";INT(A(A(I,R+1)*1000+.5)/1000
430   NEXT I
440   END
```

APPENDIX
B

MATRICES

A matrix is a rectangular array of numbers having m horizontal rows and n vertical columns. A matrix may be written in the form

$$A = \begin{vmatrix} a_{11} & a_{12} & a_{13} & \cdots & a_{1n} \\ a_{21} & a_{22} & a_{23} & \cdots & a_{2n} \\ a_{31} & a_{32} & a_{33} & \cdots & a_{3n} \\ \cdots & \cdots & \cdots & \cdots & \cdots \\ a_{m1} & a_{m2} & a_{m3} & \cdots & a_{mn} \end{vmatrix}$$

The order of a matrix is $m \times n$ or m by n. Each number in a matrix is called an *element* of the matrix. The subscript j is used to designate a row of the matrix, and the subscript k is used to designate a column of the matrix. A matrix is often designated by the symbol a_{jk}.

A matrix having only one row is known as a *row matrix* or a *row vector*. A matrix having only one column is known as a *column matrix* or a *column vector*. If the number of rows and columns is equal, the matrix is known as a *square matrix*.

The *transpose* of a matrix is the interchanging of its rows and columns and is generally designated with a superscript T. Thus $\mathbf{A}^T$ is the transpose of $\mathbf{A}$.

316

$$\mathbf{A} = \begin{vmatrix} 1 & 3 & 5 \\ 2 & 7 & 9 \\ 6 & 8 & 4 \end{vmatrix} \qquad \mathbf{A}^T = \begin{vmatrix} 1 & 2 & 6 \\ 3 & 7 & 8 \\ 5 & 9 & 4 \end{vmatrix}$$

A *minor* of a matrix is formed by removing all elements of a row and column thus forming another matrix of order $(n-1)$. The minor corresponding to the element 9 in the third column and second row of the matrix **A** above is

$$\begin{vmatrix} 1 & 3 & 5 \\ 2 & 7 & 9 \\ 6 & 8 & 4 \end{vmatrix} = \begin{vmatrix} 1 & 3 \\ 6 & 8 \end{vmatrix}$$

Multiplication of the minor of a_{jk} by $(-1)^{j+k}$ produces the cofactor of a_{jk} which is generally denoted by A_{jk}. The cofactor corresponding to element 3 in matrix **A** above will be

$$A_{12} = (-1)^{(1+2)} \begin{vmatrix} 1 & 5 \\ 6 & 4 \end{vmatrix} = - \begin{vmatrix} 1 & 5 \\ 6 & 4 \end{vmatrix} = 26$$

ADDITION AND SUBTRACTION

If two matrices have the same order $\mathbf{A} = a_{jk}$ and $\mathbf{B} = b_{jk}$, the sum or difference of the two matrices becomes

$$\mathbf{A} = \begin{vmatrix} 1 & 3 & 5 \\ 2 & 7 & 9 \\ 6 & 8 & 4 \end{vmatrix} \qquad \mathbf{B} = \begin{vmatrix} 1 & 2 & 6 \\ 3 & 7 & 8 \\ 5 & 9 & 4 \end{vmatrix}$$

$$\mathbf{A} \pm \mathbf{B} = a_{jk} \pm b_{jk}$$

$$\mathbf{A} + \mathbf{B} = \begin{vmatrix} 1+1 & 3+2 & 5+6 \\ 2+3 & 7+7 & 9+8 \\ 6+5 & 8+9 & 4+4 \end{vmatrix} = \begin{vmatrix} 2 & 5 & 11 \\ 5 & 14 & 17 \\ 11 & 17 & 8 \end{vmatrix}$$

$$\mathbf{A} - \mathbf{B} = \begin{vmatrix} 1-1 & 3-2 & 5-6 \\ 2-3 & 7-7 & 9-8 \\ 6-5 & 8-9 & 4-4 \end{vmatrix} = \begin{vmatrix} 0 & 1 & -1 \\ -1 & 0 & 1 \\ 1 & -1 & 0 \end{vmatrix}$$

MULTIPLICATION

If a matrix is multiplied by a scalar number r, the product is defined as $r\mathbf{A} = ra_{jk}$. The product of matrix **A** above and the scalar number 5 becomes

$$5\mathbf{A} = 5 \begin{vmatrix} 1 & 3 & 5 \\ 2 & 7 & 9 \\ 6 & 8 & 4 \end{vmatrix} = \begin{vmatrix} 5 & 15 & 25 \\ 10 & 35 & 45 \\ 30 & 40 & 20 \end{vmatrix}$$

If matrix **C** (c_{jk}) is multiplied by matrix **D** (d_{jk}), matrix **E** (e_{jk}) results.

$$e_{jk} = \sum_{l=1}^{n} c_{jl}d_{lk}$$

Thus if

$$C = \begin{vmatrix} 2 & 4 & 6 \\ 3 & 5 & 9 \\ 7 & 8 & 1 \end{vmatrix} \quad \text{and} \quad D = \begin{vmatrix} 7 & 9 & 8 \\ 1 & 3 & 5 \\ 2 & 6 & 4 \end{vmatrix}$$

$$E = CD$$

$$= \begin{vmatrix} 2\times7+4\times1+6\times2 & 2\times9+4\times3+6\times6 & 2\times8+4\times5+6\times4 \\ 3\times7+5\times1+9\times2 & 3\times9+5\times3+9\times6 & 3\times8+5\times5+9\times4 \\ 7\times7+8\times1+1\times2 & 7\times9+8\times3+1\times6 & 7\times8+8\times5+1\times4 \end{vmatrix}$$

$$E = \begin{vmatrix} 30 & 66 & 60 \\ 44 & 96 & 85 \\ 59 & 93 & 100 \end{vmatrix}$$

INVERSE

If **A** is a nonsingular square matrix (its determinant is not zero), there exists a unique inverse of **A** or A^{-1} such that when a matrix is multiplied by its inverse an identity matrix or unit matrix results. The identity matrix is to matrix algebra as the number 1 is to ordinary algebra.

$$AA^{-1} = I$$

The inverse of a matrix is formed by dividing the transposed matrix of cofactors by the determinant of the original matrix.

$$A^{-1} = \frac{(A_{jk})^{T}}{\det(A)}$$

Example. Invert the matrix

$$A = \begin{vmatrix} 1 & 2 & -3 \\ 2 & -2 & 1 \\ -1 & 3 & 2 \end{vmatrix}$$

Solution

$$\det(A) = -4 - 18 - 2 + 6 - 8 - 3 = -29$$

Compute the cofactor matrix and its transform:

$$A_{11} = -1^{2} \begin{vmatrix} -2 & 1 \\ 3 & 2 \end{vmatrix} = -7 \qquad A_{12} = -1^{3} \begin{vmatrix} 2 & 1 \\ -1 & 2 \end{vmatrix} = -5$$

$$A_{13} = -1^{4} \begin{vmatrix} 2 & -2 \\ -1 & 3 \end{vmatrix} = 4 \qquad A_{21} = -1^{3} \begin{vmatrix} 2 & -3 \\ 3 & 2 \end{vmatrix} = -13$$

$$A_{22} = -1^{4} \begin{vmatrix} 1 & -3 \\ -1 & 2 \end{vmatrix} = -1 \qquad A_{23} = -1^{5} \begin{vmatrix} 1 & 2 \\ -1 & 3 \end{vmatrix} = -5$$

$$\mathbf{A}_{31} = -1^4 \begin{vmatrix} 2 & -3 \\ -2 & 1 \end{vmatrix} = -4 \qquad \mathbf{A}_{32} = -1^5 \begin{vmatrix} 1 & -3 \\ 2 & 1 \end{vmatrix} = -7$$

$$\mathbf{A}_{33} = -1^6 \begin{vmatrix} 1 & 2 \\ 2 & -2 \end{vmatrix} = -6$$

$$\mathbf{A}_{jk} = \begin{vmatrix} -7 & -5 & 4 \\ -13 & -1 & -5 \\ -4 & -7 & -6 \end{vmatrix} \qquad (\mathbf{A}_{jk})^T = \begin{vmatrix} -7 & -13 & -4 \\ -3 & -1 & -7 \\ 4 & -5 & -6 \end{vmatrix}$$

and finally the inverted matrix becomes

$$\mathbf{A}^{-1} = \frac{-1}{29} \begin{vmatrix} -7 & -13 & -4 \\ -5 & -1 & -7 \\ 4 & -5 & -6 \end{vmatrix} = \begin{vmatrix} 7/29 & 13/29 & 4/29 \\ 5/29 & 1/29 & 7/29 \\ -4/29 & 5/29 & 6/29 \end{vmatrix}$$

As a check

$$\mathbf{A}\mathbf{A}^{-1} = \begin{vmatrix} 1 & 2 & -3 \\ 2 & -2 & 1 \\ -1 & 3 & 2 \end{vmatrix} \begin{vmatrix} 7/29 & 13/29 & 4/29 \\ 5/29 & 1/29 & 1/29 \\ -4/29 & 5/29 & 6/29 \end{vmatrix} = \begin{vmatrix} 1 & 0 & 0 \\ 0 & 1 & 0 \\ 0 & 0 & 1 \end{vmatrix} = \mathbf{I}$$

The following is a computer program which will perform a matrix inversion for up to a 10×10 matrix.

```
10    PRINT "MATRIX INVERSION"

20    PRINT "THIS PROGRAM WILL INVERT SQUARE MATRICES ONLY"

30    DIM A(10,10), B(10,10)

40    INPUT "DIMENSION OF MATRIX";R

50    PRINT "MATRIX ELEMENTS"

60    FOR J = 1 TO R

70    PRINT "ROW";J

80    FOR I = 1 TO R

90    PRINT "VALUE COLUMN";I;

100   INPUT A(J,I)

110   NEXT I

120   B(J,J)=1

130   NEXT J

140   FOR J = 1 TO R

150   FOR I = J TO R

160   IF A(I,J)<>0 THEN 200

170   NEXT I

180   PRINT "SINGULAR MATRIX"
```

```
190    GO TO 490

200    FOR K = 1 TO R

210    S = A(J,K)

220    A(J,K) = A(I,K)

230    A(I,K) = S

240    S = B(J,K)

250    B(J,K) = B(I,K)

260    B(I,K) = S

270    NEXT K

280    T = 1/A(J,J)

290    FOR K = 1 TO R

300    A(J,K) = T*A(J,K)

310    B(J,K) = T*B(J,K)

320    NEXT K

330    FOR L = 1 TO R

340    IF L = J THEN 400

350    T = -A(L,J)

360    FOR K = 1 TO R

370    A(L,K) = A(L,K) + T*A(J,K)

380    B(L,K) = B(L,K) + T*B(J,K)

390    NEXT K

400    NEXT L

410    NEXT J

420    PRINT

430    FOR I = 1 TO R

440    FOR J = 1 TO R

450    PRINT INT(B(I,J)*1000 + .5)/1000;" ";

460    NEXT J

470    PRINT

480    NEXT I

490    END
```

APPENDIX
C

CONVERSION FACTORS

Conventional and International System Units Conversion Factors

Distance
1 ft = 0.3048 m
1 in = 25.4 mm
1 m = 39.372 in

Force
$1 \, \text{lb}_f = 4.448 \, \text{N}$
$1 \, \text{N} = 0.2248 \, \text{lb}_f$

Mass
$1 \, \text{slug} = 1 \, \text{lb}/(\text{ft} \cdot \text{s}^2)$
1 slug = 14.59 kg
$1 \, \text{lb}_m = 0.03108 \, \text{slug}$
1 kg = 0.6852 slug

Mass moment of inertia
$1 \, \text{slug} \cdot \text{ft}^2 = 1.356 \, \text{kg} \cdot \text{m}^2$
$1 \, \text{kg} \cdot \text{m}^2 = 0.7376 \, \text{slug} \cdot \text{ft}^2$

Other
$1 \, \text{lb}_f \cdot \text{in} = 0.11298 \, \text{N} \cdot \text{m}$
$1 \, \text{mi}/\text{h} = 1.61 \, \text{km}/\text{h}$
$1 \, \text{lb}_f \cdot \text{ft} = 1.3557 \, \text{N} \cdot \text{m}$
$1 \, \text{lb}/\text{ft}^2 = 6894.7 \, \text{N}/\text{m}^2$

APPENDIX
D

INVOLUTE FUNCTIONS

Table of involute functions

Angle	0.0	0.2	0.4	0.6	0.8
14	0.004982	0.005202	0.005429	0.005662	0.005903
15	0.006150	0.006404	0.006665	0.006934	0.007350
16	0.007493	0.007784	0.008082	0.008388	0.008702
17	0.009025	0.009355	0.009694	0.010041	0.010396
18	0.010760	0.011133	0.011515	0.011906	0.012306
19	0.012715	0.013134	0.013562	0.013999	0.014447
20	0.014904	0.015372	0.015850	0.016337	0.016836
21	0.017345	0.017865	0.018395	0.018937	0.019490
22	0.020054	0.020630	0.021216	0.021815	0.022426
23	0.020054	0.020630	0.021216	0.021815	0.022426
24	0.026350	0.027048	0.027760	0.028485	0.029223
25	0.029975	0.030741	0.031521	0.032315	0.033124
26	0.033947	0.034785	0.035637	0.036505	0.037388
27	0.038287	0.039201	0.040131	0.041076	0.042039
28	0.043017	0.044012	0.045024	0.046054	0.047100
29	0.048164	0.049245	0.050344	0.051462	0.052597
30	0.053751	0.054924	0.056116	0.057267	0.058558
31	0.059809	0.061079	0.062369	0.063680	0.065012
32	0.066364	0.067738	0.069133	0.070549	0.071988
33	0.073449	0.074932	0.076439	0.077968	0.079520
34	0.081097	0.082697	0.084321	0.085970	0.087644
35	0.089342	0.091066	0.092816	0.094592	0.096395

INDEX